Barker • Rogers • Van Dyke

Student's Solutions Manual to accompany

Basic Algebra

fourth edition

Cheryl Roberts
*Northern Virginia Community College-
Manassas Campus*

SAUNDERS COLLEGE PUBLISHING
Harcourt Brace College Publishers

Fort Worth • Philadelphia • San Diego • New York • Orlando • Austin
San Antonio • Toronto • Montreal • London • Sydney • Tokyo

Copyright © 1995 by Harcourt Brace & Company

All rights reserved. No part of this publication may be reproduced or transmitted in any form or by any means, electronic or mechanical, including photocopy, recording, or any information storage and retrieval system, without permission in writing from the publisher.

Requests for permission to make copies of any part of the work should be mailed to: Permissions Department, Harcourt Brace & Company, 8th Floor, Orlando, Florida 32887-6777.

Printed in the United States of America.

Roberts: Student's Solutions Manual to accompany <u>Basic Algebra</u>, 4/e by Barker/Rogers/Van Dyke

ISBN 0-03-010243-X

6 022 98765432

CONTENTS

Chapter 0	Arithmetic Review	
Exercises 0.1	Fractions	1
Exercises 0.2	Decimals	4
Exercises 0.3	Percents	7
Exercises 0.4	Exponents and Roots	8
Exercises 0.5	Order of Operations	9
Chapter 0	Concept Review	11
Chapter 0	Test	12
Chapter 1	The Language of Algebra: Signed Numbers	
Exercises 1.1	The Language of Algebra	16
Exercises 1.2	Opposites and Absolute Value	18
Exercises 1.3	Adding of Signed Numbers	20
Exercises 1.4	Subtracting Signed Numbers	22
Exercises 1.5	Multiplying and Dividing Signed Numbers	24
Exercises 1.6	Order of Operations	26
Chapter 1	Concept Review	30
Chapter 1	Test	31
Chapter 2	Linear Equations and Inequalities	
Exercises 2.1	Simplifying Algebraic Expressions	34
Exercises 2.2	Solving Linear Equations by Adding and Subtracting	35
Exercises 2.3	Solving Linear Equations by Multiplying and Dividing	37
Exercises 2.4	Solving Linear Equations	41
Exercises 2.5	Solving Percent Problems	45
Exercises 2.6	Solving and Graphing Inequalities	48
Chapter 2	Concept Review	53
Chapter 2	Test	54
Chapter 3	Algebra of Polynomials	
Exercises 3.1	Whole-Number Exponents	59
Exercises 3.2	Integer Exponents	61
Exercises 3.3	Scientific Notation	65
Exercises 3.4	Introduction to Polynomials	68
Exercises 3.5	Adding and Subtracting Polynomials	71
Exercises 3.6	Applications	74
Exercises 3.7	Multiplying Polynomials	76
Exercises 3.8	Multiplying Binomials	80
Exercises 3.9	Special Binomial Products	83
Exercises 3.10	Dividing Polynomials	85
Chapter 3	Concept Review	89
Chapter 3	Test	91
Chapter 4	Factoring Polynomials	
Exercises 4.1	Factors and the Zero-Product Property	95
Exercises 4.2	Monomial Factors	97
Exercises 4.3	Factoring by Grouping	100
Exercises 4.4	Factoring Trinomials of the Form $x^2 + bx + c$	103
Exercises 4.5	Factoring Trinomials of the Form $ax^2 + bx + c$	106
Exercises 4.6	Special Cases	111
Exercises 4.7	Factoring Polynomials: A Review	114
Chapter 4	Concept Review	117
Chapter 4	Test	118

Chapter 5	**Fractions of Algebra**	
Exercises 5.1	Multiplying and Reducing Rational Expressions	121
Exercises 5.2	Dividing Rational Expressions	124
Exercises 5.3	Least Common Multiple: Solving Rational Equations	128
Exercises 5.4	Adding and Subtracting Rational Expressions	132
Exercises 5.5	Complex Fractions	139
Exercises 5.6	Ratio and Proportion	144
Exercises 5.7	Variation	148
Chapter 5	Concept Review	151
Chapter 5	Test	152
Chapter 6	**Linear Equations in Two Variables**	
Exercises 6.1	The Rectangular Coordinate System	157
Exercises 6.2	Solutions and Graphs	160
Exercises 6.3	Slopes and Intercepts	166
Exercises 6.4	Graphing Using Slope and Intercept	170
Exercises 6.5	Equations of Lines	176
Exercises 6.6	Graphing of Linear Inequalities	180
Chapter 6	Concept Review	183
Chapter 6	Test	184
Chapter 7	**Systems of Equations**	
Exercises 7.1	Solving Systems by Graphing	188
Exercises 7.2	Solving Systems by Substitution	193
Exercises 7.3	Solving Systems by Addition	202
Exercises 7.4	Applications of Systems: A Review	208
Chapter 7	Concept Review	210
Chapter 7	Test	210
Chapter 8	**Square Roots and Related Equations**	
Exercises 8.1	Square Roots	214
Exercises 8.2	Simplifying Radical Expressions	216
Exercises 8.3	Multiplying Radicals	218
Exercises 8.4	Dividing Radicals	221
Exercises 8.5	Adding and Subtracting Radicals	226
Exercises 8.6	Solving Radical Equations	229
Chapter 8	Concept Review	235
Chapter 8	Test	236
Chapter 9	**Quadratic Equations**	
Exercises 9.1	Factoring Revisited: The Square Root Property of Equality	239
Exercises 9.2	The Pythagorean Theorem	241
Exercises 9.3	Completing the Square	244
Exercises 9.4	The Quadratic Formula	251
Exercises 9.5	Quadratic Equations: A Review	260
Exercises 9.6	Graphing Parabolas	267
Chapter 9	Concept Review	271
Chapter 9	Test	271

PREFACE

This *Student Solutions Manual* accompanies *Basic Algebra, Fourth Edition* by Barker/Rogers/VanDyke. The Manual contains solutions to every other odd-numbered exercise, beginning with number 1 in the section exercise sets and every exercise in the Chapter Concept Review and Chapter Test. I have made every effort to prepare the Manual error-free. If you discover any typographical or mathematical errors or have suggestions for improvement, please contact me at Northern Virginia Community College, 6901 Sudley Road, Manassas, VA 22110.

Acknowledgements I give special thanks to Nancy Varner, my mother, for the many long hours spent at the computer typing this manual.

Cheryl V. Roberts

CHAPTER 0

EXERCISES 0.1

1. $\dfrac{5}{6}, \dfrac{11}{15}, \dfrac{3}{8}$ The numerator is smaller than the denominator.

5. $\dfrac{4}{8} = \dfrac{4 \cdot 1}{4 \cdot 2}$ Divide out the greatest common factor of 4.

    $= \dfrac{1}{2}$

9. $\dfrac{2}{5} \cdot \dfrac{3}{8} = \dfrac{1}{5} \cdot \dfrac{3}{4}$ 2 and 8 have a common factor of 2.

    $= \dfrac{3}{20}$ Multiply.

13. $\dfrac{1}{2} + \dfrac{1}{6} = \dfrac{1 \cdot 3}{2 \cdot 3} + \dfrac{1}{6}$ The common denominator is 6.

    $= \dfrac{3}{6} + \dfrac{1}{6}$ Write each fraction with a denominator of 6.

    $= \dfrac{4}{6}$ Add.

    $= \dfrac{2}{3}$ Reduce.

17. $\dfrac{24}{36} = \dfrac{12 \cdot 2}{12 \cdot 3} = \dfrac{2}{3}$ Divide out the greatest common factor of 12.

21. $\dfrac{12}{21} \cdot \dfrac{14}{24} = \dfrac{1 \cdot 2}{3 \cdot 2}$ 12 and 24 have a common factor of 12.

                     14 and 21 have a common factor of 7.

    $= \dfrac{1 \cdot 1}{3 \cdot 1}$ 2 and 2 have a common factor of 2.

    $= \dfrac{1}{3}$ Multiply.

25. $\dfrac{13}{18} + \dfrac{7}{27} = \dfrac{13 \cdot 3}{18 \cdot 3} + \dfrac{7 \cdot 2}{27 \cdot 2}$ The common denominator is 54.

    $= \dfrac{39}{54} + \dfrac{14}{54}$ Write each fraction with a denominator of 54.

    $= \dfrac{53}{54}$ Add.

29. $\dfrac{96}{144} = \dfrac{48 \cdot 2}{48 \cdot 3}$ Divide out the greatest common factor of 48.

$= \dfrac{2}{3}$

33. $\left(2\dfrac{7}{8}\right)\left(1\dfrac{1}{3}\right) = \left(\dfrac{23}{8}\right)\left(\dfrac{4}{3}\right)$ Change to improper fractions.

$= \dfrac{23 \cdot 1}{2 \cdot 3}$ 8 and 4 have a common factor of 4.

$= \dfrac{23}{6}$ Multiply.

$= 3\dfrac{5}{6}$ Change to a mixed number.

37. $5\dfrac{2}{9} + 3\dfrac{5}{6} = 5\dfrac{2 \cdot 2}{9 \cdot 2} + 3\dfrac{5 \cdot 3}{6 \cdot 3}$ 1 and 4 have a common denominator of 4.

$= 5\dfrac{4}{18} + 3\dfrac{15}{18}$ Subtract.

$= 8\dfrac{19}{18}$ 22 and 4 have a common factor of 2.

$= 8 + 1\dfrac{1}{18}$ Change to a mixed number.

Add.

$= 9\dfrac{1}{18}$

41. $1 - \dfrac{1}{4} = \dfrac{4}{4} - \dfrac{1}{4}$ The common denominator is 18.

Write each fraction with a denominator of 18.

$= \dfrac{3}{4}$

$(22)\left(\dfrac{3}{4}\right) = \left(\dfrac{11}{1}\right)\left(\dfrac{3}{2}\right)$ Add whole numbers and add fractions.

$= \dfrac{33}{2}$ Change $\dfrac{19}{18}$ to a mixed number.

$= 16\dfrac{1}{2}$ Add.

$16\dfrac{1}{2}$ gal

Exercises 0.1

45. $\dfrac{7}{8} + \dfrac{3}{16} + \dfrac{1}{2} + \dfrac{1}{8} + \dfrac{1}{4}$

$= \dfrac{7 \cdot 2}{8 \cdot 2} + \dfrac{3}{16} + \dfrac{1 \cdot 8}{2 \cdot 8} + \dfrac{1 \cdot 2}{8 \cdot 2} + \dfrac{1 \cdot 4}{4 \cdot 4}$ The common denominator is 16.

$= \dfrac{14}{16} + \dfrac{3}{16} + \dfrac{8}{16} + \dfrac{2}{16} + \dfrac{4}{16}$ Write each fraction with a denominator of 16.

$= \dfrac{31}{16}$ Add.

$= 1\dfrac{15}{16}$ Change to a mixed number.

$1\dfrac{15}{16}$ in.

49. $9\dfrac{3}{16} \div 5 = \dfrac{147}{16} \div \dfrac{5}{1}$ Change to improper fractions.

$= \dfrac{147}{16} \cdot \dfrac{1}{5}$ Invert and mutliply.

$= \dfrac{147}{80}$

$= 1\dfrac{67}{80}$

$1\dfrac{67}{80}$ cups of nuts

$8\dfrac{1}{6} \div 5 = \dfrac{49}{6} \div \dfrac{5}{1}$ Change to improper fractions.

$= \dfrac{49}{6} \cdot \dfrac{1}{5}$ Invert and mutliply.

$= \dfrac{49}{30}$

$= 1\dfrac{19}{30}$

$1\dfrac{19}{30}$ cups of raisins

53. $\dfrac{5}{8} \cdot \dfrac{4}{5} - \dfrac{7}{8} \cdot \dfrac{1}{2} = \dfrac{1}{2} \cdot \dfrac{1}{1} - \dfrac{7}{8} \cdot \dfrac{1}{2}$ 5 and 5 have a common factor of 5.
   4 and 8 have a common factor of 4.

$= \dfrac{1}{2} - \dfrac{7}{16}$ Multiply.

$= \dfrac{1 \cdot 8}{2 \cdot 8} - \dfrac{7}{16}$ The common denominator is 16.

$= \dfrac{8}{16} - \dfrac{7}{16}$ Write each fraction with a denominator of 16.

$= \dfrac{1}{16}$ Subtract.

EXERCISES 0.2

1. 0.5
    0.6

    1.1 $5 + 6 = 11$, carry 1.

5. 3.9
    1.8

    2.1   Subtract and align decimal.

9. 1.8
    0.4

    0.72   There is a total of 2 decimal places.

13. $0.5 \overline{)1.2\,55}$ with quotient 2.51
    10
    --
    25
    25
    --
    5
    5
    -
    0   Move each decimal 1 place to the right.

17. $2\overline{)1.0}$ with quotient 0.5
    10
    --
    0   Add one zero.
    2 doesn't divide 1.

21. $0.24 = \dfrac{24}{100}$ Read twenty-four hundredths.

    $= \dfrac{6}{25}$ Reduce.

Exercises 0.2

25. 3.500   Write two zeros.
    2.850   Write one zero.
    1.891
    0.340   Write one zero.
    8.581   Add.

29. 9 11 12
    $\not{1}\not{0}1.\not{2}\not{3}13$
    102.33
     92.45
      9.88

Borrow 1 tenth from the 3 in the tenths place to add to the 3 in the hundredths place. 1 tenth = 10 hundredths.

Borrow 1 from the 2 in the units place to add to the 2 in the tenths place. 1 unit = 10 tenths.

Borrow 1 from the 1 in the hundreds place, borrow 1 from the 10 in the tens place.

33.   2.54
     1.34
    1016
     762
     254
   3.4036   There is a total of four decimal places.

37.   2.351
   15$\overline{)35.265}$
       30
        52
        45
        76
        75
        15
        15
         0

41.   0.55
   20$\overline{)11.00}$
     100
     100
     100
       0

Divide the numerator by the denominator.
(11 = 11.00) Insert zeros to complete the division.

45. $0.85 = \dfrac{85}{100}$

Read as eighty-five hundredths. Reduce by dividing out a common factor of 5.

$ = \dfrac{17}{20}$

49.   198.7614
    13.0200
     8.0004
     0.7640
+ 142.6200
   363.1658

Write 13.02 as 13.0200.
Write 0.764 as 0.7640.
Write 142.62 as 142.6200.

Add.

53.
```
       15 11
    5  5̸1 10
   1̸6̸.6̸2̸0̸
 -   3.821
   12.799
```

Write 16.62 as 16.620

Borrow 1 hundredth from the 2 in the hundredths place to add to the 0 in the thousandths place.
1 hundredth = 10 thousandths.

Borrow 1 tenth from the 6 in the tenths place to add to the 1 in the hundredths place.
1 tenth = 10 hundredths.

Borrow 1 unit from the 6 in the units place to add to the 5 in the tenths place. 1 unit = 10 tenths.

57.
```
     14.09
   × 23.12
     2818
     1409
     4227
    2818
   325.7608
```

There is a total of 4 decimal places.

61.
```
           29.9
   0.016 )0.478.4
           32
           158
           144
            144
            144
              0
```

Move both decimals 3 places to the right.

65.
```
         0.916
     12 )11.000
         10 8
            20
            12
            80
            72
             8
```

Divide the numerator by the denominator.
(11 = 11.000) Insert zeros.

12 doesn't divide 11 but it will divide 110.

0.916 rounds to 0.92 to the nearest hundredth.

Align decimals and add.

69.
```
      8.6
     14.9
     15.4
   + 13.5
     52.4
```

52.4 gal

73.
```
         0.8125
     16 )13.0000
         12 8
            20
            16
            40
            32
            80
            80
             0
```

Divide the numerator by the denominator.
(13 = 13.0000) Insert zeros.

Exercises 0.3

77. One possible answer: Before adding you must determine the least common denominator. Then change each fraction to an equivalent fraction with the LCD as denominator. Next add the fractions by adding the numerators and keep the LCD as denominator. Reduce if possible.

81. $0.87 + 0.98 - 0.12 - 0.22$
 $= 1.85 - 0.12 - 0.22$
 $= 1.73 - 0.22$
 $= 1.51$

Add and subtract from left to right.

EXERCISES 0.3

1. $\dfrac{1}{10} = 0.1$

 Divide the numerator 1 by the denominator 10.

 $0.1 = 10\%$

 Move the decimal 2 places to the right.

5. $50\% = 0.5$

 Move the decimal 2 places to the left.

 $0.5 = \dfrac{5}{10} = \dfrac{1}{2}$

 0.5 is read as "five tenths".

9. $0.3125 = 31.25\%$

 Move the decimal 2 places to the right.

 $0.3125 = \dfrac{3125}{10000}$

 0.3125 is read as "three thousand one hundred twenty-five ten thousandths".

 $= \dfrac{5}{16}$

13. $\dfrac{7}{10} = 0.7$

 Divide the numerator 7 by the denominator 10.

 $0.7 = 70\%$

 Move the decimal 2 places to the right.

17. $\dfrac{1}{32} = 0.03125$

 Divide the numerator 1 by the denominator 32.

 $0.03125 = 3.125\%$

 Move the decimal 2 places to the right.

21. $3.45 = 345\%$

 Move the decimal 2 places to the right.

 $3.45 = 3\dfrac{45}{100}$

 3.45 is read as "three and forty-five hundredths".

 $= 3\dfrac{9}{20}$

25. $0.89 = 89\%$

 Move the decimal 2 places to the right.

29. $\dfrac{0.22}{100} = 0.22\%$

 Percent is based on a denominator of 100.

33. One possible answer: First change the fraction to a decimal by dividing the numerator by the denominator. Then change the decimal to a percent by moving the decimal 2 places to the right and inserting a percent symbol.

37. $16\frac{1}{4}\% = \dfrac{16\frac{1}{4}}{100}$ Percent is based on a denominator of 100.

$= \dfrac{\frac{65}{4}}{100}$ Change $16\frac{1}{4}$ to an improper fraction

$= \dfrac{65}{4} \div \dfrac{100}{1}$

$= \dfrac{65}{4} \cdot \dfrac{1}{100}$ Invert and multiply.

$= \dfrac{13}{80}$

EXERCISES 0.4

1. $6^2 = 6 \cdot 6$
    $= 36$ There are 2 factors of 6.

5. $10^2 = 10 \cdot 10$
    $= 100$ There are 2 factors of 10.

9. $5 \cdot 5 \cdot 5 \cdot 5 = 5^4$ There are 4 factors of 5.

13. $\sqrt{25} = 5$ $5^2 = 25$.

17. $\sqrt{81} = 9$ $9^2 = 81$.

21. $6^3 = 6 \cdot 6 \cdot 6$
     $= 216$ There are 3 factors of 6.

25. $2 \cdot 2 \cdot 2 \cdot 2 \cdot 2 \cdot 2 = 2^6$ There are 6 factors of 2.

29. $6 \cdot 6 \cdot 6 \cdot 6 \cdot 6 \cdot 6 \cdot 6 = 6^7$ There are 7 factors of 6.

33. $\sqrt{225} = 15$ $15^2 = 225$.

37. $11^3 = 11 \cdot 11 \cdot 11$
     $= 1331$ There are 3 factors of 11.

41. $5^4 = 5 \cdot 5 \cdot 5 \cdot 5$
     $= 625$ There are 4 factors of 5.

Section 0.5

45. $2 \cdot 2 \cdot 3 \cdot 3 = 2^2 \cdot 3^2$ There are 2 factors of 2 and 2 factors of 3.

49. $\sqrt{484} = 22$ $22^2 = 484$.

53. $4 \times 10^6 = 4000000$ Move the decimal 6 places to the right.
$4,000,000$

57. $a^2 + b^2 = c^2$ Replace the shorter sides a and b by 10 and 24.
$10^2 + 24^2 = c^2$
$100 + 576 = c^2$
$676 = c^2$
$\sqrt{676} = \sqrt{c^2}$
$26 = c$ $26^2 = 676$.

61. $\sqrt{65536} = \sqrt{4^8}$ Factor 65536.
$= \sqrt{(4^4)^2}$ Write 4^8 as a perfect square.
$= 4^4$ or 256

65. $5^3 \cdot 3^2 = x$ $5^3 = 125$.
$125 \cdot 9 = x$ $3^2 = 9$.
$1125 = x$ Multiply.

EXERCISES 0.5

1. $3 + 3 \cdot 4 = 3 + 12$ Multiply first.
$= 15$ Add.

5. $15 - 2 \cdot 6 = 15 - 12$ Multiply first.
$= 3$ Subtract.

9. $7 \cdot 2 - 8 \div 2 = 14 - 8 \div 2$ Multiply first before dividing since it appears on the left.
$= 14 - 4$ Divide next.
$= 10$ Subtract last.

13. $5^2 + 4 \cdot 2 = 25 + 4 \cdot 2$ Exponents first. ($5^2 = 25$)
$= 25 + 8$ Multiply next. ($4 \cdot 2 = 8$)
$= 33$ Add.

17. $5 \cdot 4 - 36 \div 9 + 12$
$= 20 - 36 \div 9 + 12$ Multiply first.
$= 20 - 4 + 12$ Divide next.
$= 16 + 12$ Add and subtract left to right.
$= 28$

21. $75 - 5^2(2)$
$= 75 - 25(2)$ Exponents first.
$= 75 - 50$ Multiply next.
$= 25$ Subtract.

25. $5(15 - 3) + 4(16 - 9)$
    $= 5(12) + 4(7)$ Paretheses first.
    $= 60 + 28$ Multiplication next.
    $= 88$ Add.

29. $\dfrac{12(9-5)}{4^2} = \dfrac{12(4)}{16}$ Parentheses first in the numerator. Exponent in denominator.

   $= \dfrac{48}{16}$ Multiply next in numerator.

   $= 3$ Divide.

33. $\dfrac{85 - 75 \div 3 \cdot 2 + 5 \cdot 4 \div 2}{9^2 - 6^2}$

   $= \dfrac{85 - 25 \cdot 2 + 5 \cdot 4 \div 2}{81 - 36}$ Divide first in numerator. Exponents in denominator.

   $= \dfrac{85 - 50 + 20 \div 2}{45}$ Multiply left to right in numerator. Subtract in denominator.

   $= \dfrac{85 - 50 + 10}{45}$ Divide in numerator.

   $= \dfrac{45}{45}$ Add and subtract left to right.

   $= 1$ Divide.

37. $(5^2 - 4^2) + 7 - (10 - 8)^3 + 75 \div 5^2$
    $= (25 - 16) + 7 - (2)^3 + 75 \div 25$ Exponents and inside parentheses first.
    $= 9 + 7 - 8 + 75 \div 25$ Subtract inside parentheses. $2^3 = 8$
    $= 9 + 7 - 8 + 3$ Divide next.
    $= 11$ Add and subtract left to right.

41. $\sqrt{81}(12 + 9) - 3^2(21 - 9)$
    $= \sqrt{81}(21) - 3^2(12)$ Inside parentheses first.
    $= 9(21) - 9(12)$ $\sqrt{81} = 9$, $3^2 = 9$.
    $= 189 - 108$ Multiply left to right.
    $= 81$ Subtract.

45. $24(225) + 34(103)$ Number of suits times price per suit added to number of jackets times price per jacket.
    $= 5400 + 3502$ Multiply left to right.
    $= 8902$ Add.

   $\$8902$

49. $12(40) + 18(45 - 40)$ Inside parentheses first.
    $= 12(40) + 18(5)$ Multiply left to right.
    $= 480 + 90$ Add.
    $= 570$

   $\$570$

53. $6(2^3 \cdot 3 - 4) \div 4 - 5$
    $= 6(8 \cdot 3 - 4) \div 4 - 5$ Exponent inside parentheses first.
    $= 6(24 - 4) \div 4 - 5$ Multiply inside parentheses next.
    $= 6(20) \div 4 - 5$ Subtract inside parentheses.
    $= 120 \div 4 - 5$ Multiply.
    $= 30 - 5$ Divide.
    $= 25$ Subtract.

CHAPTER 0   CONCEPT REVIEW

1. False; 1 can be written as an improper fraction $\frac{1}{1}$. The numerator will always equal the denominator so it cannot be written as a proper fraction.

2. True

3. False; $\frac{3}{2}$ is an improper fraction and it cannot be reduced further.

4. True

5. True

6. True

7. True

8. False; $8.5 - 0.3 = 8.2$

9. False; All fractions can be changed to exact decimals or repeating decimals.

10. True

11. True

12. True

13. False; To change a decimal to a percent move the decimal point two places to the right and insert a percent symbol.

14. True

15. True

16. True

17. False; 4 is the base and 3 is the exponent.

18. False; 3 is the base.

19. True

20. False; In the expression $(3 + 4)^2$, addition takes precedence over exponents.

21. False; In the expression $3(5 - 1)$, subtraction takes precedence over multiplication.

22. True

CHAPTER 0 TEST

1. $9^3 = 9 \cdot 9 \cdot 9$
    $= 729$

    9 is a factor 3 times.

2. $48\% = 0.48$

    Move the decimal 2 places to the left and drop the % symbol.

3.   0.760
       9.200
       8.098
    $+\ \underline{3.000}$
      21.058

    Write 0.76 as 0.760.
    Write 9.2 as 9.200.

    Write 3 as 3.000.
    Add.

4. $72\% = \dfrac{72}{100}$

    $= \dfrac{18}{25}$

    Percent is based on a denominator of 100.

    Reduce.

5. $\dfrac{28}{84} = \dfrac{28 \cdot 1}{28 \cdot 3}$

    $= \dfrac{1}{3}$

    Divide out a common factor of 28.

6. $\sqrt{225} = 15$

    $15^2 = 225$.

7. $\dfrac{5}{8} + \dfrac{4}{5} + \dfrac{3}{4} = \dfrac{5 \cdot 5}{8 \cdot 5} + \dfrac{4 \cdot 8}{5 \cdot 8} + \dfrac{3 \cdot 10}{4 \cdot 10}$

    The LCD is 40.

    $= \dfrac{25}{40} + \dfrac{32}{40} + \dfrac{30}{40}$

    Convert to equivalent fractions with denominator 40.

    $= \dfrac{25 + 32 + 30}{40}$

    Add.

    $= \dfrac{87}{40}$

    $= 2\dfrac{7}{40}$

8. $8 \cdot 8 \cdot 8 \cdot 8 \cdot 8 \cdot 8 = 8^6$

    8 is a factor 6 times.

9.   0.851
       $\underline{\ 2.76}$
       5106
       5957
      $\underline{1702\ \ }$
      2.34876

    There is a total of 5 decimal places.

Chapter 0 Test

10. $\dfrac{7}{8} = 0.875$

 $= 87.5\%$

Convert to a decimal by dividing the numerator by the denominator.

Move the decimal 2 places to the right.

11. $\dfrac{15}{16} \div \dfrac{25}{48} = \dfrac{15}{16} \cdot \dfrac{48}{25}$

$= \dfrac{3 \cdot 3}{1 \cdot 5}$

$= \dfrac{9}{5}$

$= 1\dfrac{4}{5}$

Invert the divisor and change to multiplication.

15 and 25 have a common factor of 5.
16 and 48 have a common factor of 16.

Multiply.

12.
$$
\begin{array}{r}
9\\
8\;\;10\;7\;10\;12\\
9\,.\,\cancel{0}\,\cancel{8}\,\cancel{0}\,\cancel{2}\\
-4\,.\,3\,2\,1\,4\\
\hline
4\,.\,7\,5\,8\,8
\end{array}
$$

Borrow 1 hundredth from the 8 in the hundredths place to add to the 0 in the thousandths place.
1 hundredth = 10 thousandths.

Borrow 1 thousandth from the 10 in the thousandths place to add to the 2 in the ten-thousandths place.
1 thousandth = 10 ten-thousandths.

Borrow 1 unit from the 9 in the units place to add to the 0 in the tenths place.

13. $\dfrac{4}{5},\ \dfrac{19}{20},\ \dfrac{3}{7},\ \dfrac{22}{25}$

The numerator is smaller than the denominator.

14.
$$
\begin{array}{r}
4.03\\
4.5\overline{)18.1\,35}\\
\underline{18\,0}\\
13\\
\underline{0}\\
135\\
\underline{135}\\
0
\end{array}
$$

Move both decimals 1 place to the right.

45 will not divide into 13 but it will divide into 135.

15. $0.251 = 25.1\%$

Move the decimal 2 places to the right and insert a % symbol.

16. $\left(\dfrac{5}{8}\right)\left(\dfrac{8}{9}\right)\left(\dfrac{3}{5}\right) = \dfrac{1 \cdot 1 \cdot 1}{1 \cdot 3 \cdot 1}$

$= \dfrac{1}{3}$

5 and 5 have a common factor of 5.

8 and 8 have a common factor of 8.

3 and 9 have a common factor of 3.

17. $\dfrac{7}{8} - \dfrac{11}{24} = \dfrac{7 \cdot 3}{8 \cdot 3} - \dfrac{11}{24}$

$= \dfrac{21}{24} - \dfrac{11}{24}$

The common denominator is 24.

Convert $\dfrac{7}{8}$ to a fraction with denominator 24.

$$= \frac{10}{24}$$ Subtract.

$$= \frac{5}{12}$$ Reduce.

18. $\sqrt{676} = 26$ $26^2 = 676$.

19. $12 \div 4 \cdot 3 - 16 \cdot 2 \div 8$
   $= 3 \cdot 3 - 16 \cdot 2 \div 8$
   $= 9 - 16 \cdot 2 \div 8$
   $= 9 - 32 \div 8$ Multiply and divide left to right.
   $= 9 - 4$ Subtract.
   $= 5$

20. $5(3^2 - 7)^2 + 2 \cdot 3^2$
   $= 5(9 - 7)^2 + 2 \cdot 3^2$ Exponent inside parentheses first.
   $= 5(2)^2 + 2 \cdot 3^2$ Subtract inside parentheses.
   $= 5(4) + 2 \cdot 9$ Exponents next.
   $= 20 + 18$ Multiply left to right.
   $= 38$ Add.

21.   15
     6 ͞3͞10
     57.6̶0̸   Write 57.6 as 57.60.
    - 2.89   Borrow 1 tenth from the 6 in the tenths
     54.71   place to add to the 0 in the hundredths place.
          1 tenth = 10 hundredths.

   Borrow 1 unit from the 7 in the units place to add to the 5 in the tenths place. 1 unit = 10 tenths.

22. $0.22 = \dfrac{22}{100}$ 0.22 is read as "twenty-two hundredths".

   $= \dfrac{11}{50}$ Reduce.

23.   9.27
      84.20   Write 84.2 as 84.20.
    + 123.50   Write 123.5 as 123.50.
     216.97   Add.

24.   0.4375
     16 ⟌ 7.0000   Divide the numerator by the denominator.
         6̲4̲
        60   (7 = 7.0000) Insert zeros.
        4̲8̲
       120
       1̲1̲2̲
         80
         8̲0̲
          0

25.
```
    1.89
  × 0.23
    567
   378
  0.4347
```
There is a total of 4 decimal places.

CHAPTER 1

EXERCISES 1.1

1. $x + y = 4 + 6$ Replace x with 4 and y with 6 respectively.
              $= 10$ Add.

5. $47 - b = 47 - 21$ Replace b with 21.
                   $= 26$ Subtract.

9. $b^2 = 21^2$ Replace b with 21.
           $= 441$ Square 21.

13. 20 less than a given number   "Less than" indicates subtraction from a number.

17. $x - 5 = 14$
     $19 - 5 = 14$ Replace x with 19.
         $14 = 14$ Subtract.
             True

    $x = 19$ is a solution.

21. $\dfrac{x}{y} = \dfrac{72}{18}$ Replace x with 72 and y with 18.
        $= 4$ Divide.

25. $10 - b = 10 - 10$ Replace b with 10.
                  $= 0$ Subtract.

29. $a - c = 50 - 25$ Replace a with 50 and c with 25.
               $= 25$ Subtract.

33. $x - y$ The first number is represented by x and
                                        the second number is represented by y.

37. $x + y + 18$ "Increased by" indicates addition.

41. $2x - 12 = 4$
     $2(8) - 12 = 4$ Replace x with 8.
        $16 - 12 = 4$ Multiply.
               $4 = 4$ Subtract.
              True

    $x = 8$ is a solution.

45. $xy = (6.15)(2.05)$ Replace x with 6.15 and y with 2.05.
             $= 12.6075$ Multiply.

49. $10z = 10(0.0123)$ Replace z with 0.0123.
               $= 0.123$ Multiply.

Exercises 1.1

57. $x^2 + 5y + z^3 = (6.15)^2 + 5(2.05) + (0.0123)^3$ Replace x with 6.15, y with 2.05 and z with 0.0123.
$ = 37.8225 + 10.25 + 0.000001860867$ Evaluate exponents and multiply.
$ = 48.07250186$ Add.

61. $9x - 8 = 4x + 27$
$9(7) - 8 = 4(7) + 27$ Replace x with 7.
$63 - 8 = 28 + 27$ Multiply.
$55 = 55$ Add.

  True

  $x = 7$ is a solution.

65. $D = rt$
$ = (50)(4)$ Replace r with 50 and t with 4.
$ = 200$ Multiply.

69. $A = wl$
$ = (12)(15)$ Replace w with 12 and l with 15.
$ = 180$ Multiply.

  180 ft^2 Area is measured with square units.

73. $t = \dfrac{D}{r}$

  $ = \dfrac{180}{45}$ Replace D with 180 and r with 45.

  $ = 4$ Divide.

  4 hrs

77. $IQ = \dfrac{100(MA)}{CA}$

  $ = \dfrac{100(32)}{25}$ Replace MA with 32 and CA with 25.

  $ = 4(32)$ Reduce, $\dfrac{100}{25} = 4$.

  $ = 128$ Multiply.

81. $A = 2(5 \cdot 12) + 2(6 \cdot 12)$ 2 sides with dimensions 5 cm by 12 cm;
  2 sides with dimensions 6 cm by 12 cm.
  $ = 120 + 144$ Multiply.
  $ = 264$ Add.

  Number of tiles $= \dfrac{264}{4}$
  $ = 66$ Divide.

85. $\dfrac{x^4 - 2x^2 + 1}{(x - 1)^2(x + 1)^2} = \dfrac{4^4 - 2(4)^2 + 1}{(4 - 1)^2(4 + 1)^2}$ Replace x with 4.

$= \dfrac{4^4 - 2(4)^2 + 1}{3^2 \cdot 5^2}$ Simplify inside parentheses.

$= \dfrac{256 - 2(16) + 1}{9 \cdot 25}$ Evaluate exponents.

$= \dfrac{256 - 32 + 1}{225}$ Multiply.

$= \dfrac{225}{225}$ Add and subtract.

$= 1$ Divide.

89. $\dfrac{4}{5} = 0.8$ Divide 4 by 5.

$0.8 = 80\%$ Move decimal 2 places right.

93. $2.45 = 245\%$ Move decimal 2 places right.

$245\% = \dfrac{245}{100}$ % is based on a denominator of 100.

$= \dfrac{49}{20}$ Divide numerator and denominator by 5 to reduce.

$= 2\dfrac{9}{20}$ Change to a mixed number.

97. $0.0145 = 1.45\%$ Move decimal two places right.

$0.0145 = \dfrac{145}{10000}$ 0.0145 is read as "one hundred forty-five ten thousandths".

$= \dfrac{29}{2000}$ Divide numerator and denominator by 5 to reduce.

EXERCISES 1.2

1. $-(8) = -8$ Since 8 is 8 units to the right of zero on the number line, the opposite of 8 is 8 units to the left of zero.

5. $-(-150) = 150$ The opposite of negative 150 is positive 150.

9. $|-9| = 9$ The number is negative, so write its opposite.

13. $-|-8| = -(8)$
    $= -8$
    $|-8| = 8.$
    $-(8) = -8.$

Exercises 1.2

17. $-(1.7) = -1.7$ The opposite of positive 1.7 is negative 1.7.

21. $|15.73| = 15.73$ The number is positive, so write the number.

25. $-|-47| = -[-(-47)]$
$ = -(47)$
$ = -47$

     $|-47| = -(-47)$.
     $-(-47) = 47$.
     $-(47) = -47$.

29. $-\left(-\dfrac{3}{4}\right) = \dfrac{3}{4}$ The opposite of negative $\dfrac{3}{4}$ is positive $\dfrac{3}{4}$.

33. $-[-(-21)] = -(21)$
$ = -21$

     $-(-21) = 21$.
     $-(21) = -21$.

37. $-[-(-51)] = -(51)$
$ = -51$

     $-(-51) = 51$.
     $-(51) = -51$.

41. $|-(-8.1)| = |8.1|$
$ = 8.1$

     $-(-8.1) = 8.1$.
     The number is positive, so write the number.

45. $-(-|48|) = -(-48)$
$ = 48$

     The number is positive, so write the number.
     $-(-48) = 48$.

49. -9 Opposite is indicated by putting a minus sign in front of the number.

53. $|8|$ $|8|$ indicates the distance between 8 and 0 on the real number line.

57. $-6\dfrac{3}{4}$ Loss indicates a negative number.

61. $-x - y$ Less indicates subtraction.

65. $-(x + y)$ Sum indicates addition and opposite is indicated by putting a minus sign in front of this sum.

73.
$$
\begin{array}{r}
977\frac{8}{24} \\
24\overline{\smash{)}23456} \\
\underline{216} \\
185 \\
\underline{168} \\
176 \\
\underline{168} \\
8
\end{array}
$$

$977\dfrac{8}{24} = 977\dfrac{1}{3}$ Reduce $\dfrac{8}{24}$ to $\dfrac{1}{3}$.

77.
```
         25.63
0.24 ) 6.15.12
       4 8
       ---
       1 35
       1 20
       ---
         151
         144
         ---
          72
          72
          --
           0
```
Move both decimals 2 places right.

EXERCISES 1.3

1. $-16 + 12 = -[\,|-16| - |12|\,]$
    $= -(16 - 12)$
    $= -4$

    The signs are unlike, so subtract their absolute values.
    Since −16 has the larger absolute value the sum is negative.

5. $-7 + 7 = |-7| - 7$
    $= 7 - 7$
    $= 0$

    The signs are unlike, so subtract their absolute values.

9. $7 + (-14) = -[\,|-14| - 7\,]$
    $= -(14 - 7)$
    $= -7$

    The signs are unlike, so subtract their absolute values.
    Since −14 has the larger absolute value the sum is negative.

13. $a = 12 + (-4)$
    $= 12 - |-4|$
    $= 12 - 4$
    $= 8$

    The signs are unlike, so subtract their absolute values.
    The sum is positive.

17. $-87 + 94 = 94 - |-87|$
    $= 94 - 87$
    $= 7$

    The signs are unlike, so subtract their absolute values.
    The sum is positive.

21. $-84 + 91 = 91 - |-84|$
    $= 91 - 84$
    $= 7$

    The signs are unlike, so subtract their absolute values.
    The sum is positive.

25. $-6.3 + 3.7 = -[\,|-6.3| - 3.7\,]$
    $= -(6.3 - 3.7)$
    $= -2.6$

    The signs are unlike, so subtract their absolute values.
    Since −6.3 has the larger absolute value the sum is negative.

29. $a = -25 + 32$
    $= 32 - |25|$
    $= 32 - 25$
    $= 7$

    The signs are unlike, so subtract their absolute values.
    The sum is positive.

33. $e = -21.1 + 16.8$
    $= -[\,|-21.1| - 16.8\,]$
    $= -(21.1 - 16.8)$
    $= -4.3$

    The signs are unlike, so subtract their absolute values.
    Since −21.1 has the larger absolute value the sum is negative.

Exercises 1.3

37. $-40 + 19 + (-36) + 21$
$= -21 + (-36) + 21$
$= -57 + 21$
$= -36$

$-40 + 19 = -21.$
$-21 + (-36) = -57.$
$-57 + 21 = -36.$

41. $-28.8 + (-18.02) + 36.5$
$= -46.82 + 36.5$
$= -10.32$

$-28.8 + (-18.02) = -46.82.$
$-46.82 + 36.5 = -10.32.$

45. $\dfrac{13}{9} + \left(-\dfrac{8}{3}\right) + \left(-\dfrac{7}{3}\right)$

$= \dfrac{13}{9} + \left(-\dfrac{24}{9}\right) + \left(-\dfrac{21}{9}\right)$

The common denominator is 9.

$= -\dfrac{11}{9} + \left(-\dfrac{21}{9}\right)$

$\dfrac{13}{9} + \left(-\dfrac{24}{9}\right) = -\dfrac{11}{9}.$

$= -\dfrac{32}{9}$

$-\dfrac{11}{9} + \left(-\dfrac{21}{9}\right) = -\dfrac{32}{9}.$

49. $x = 0.5 + (-0.75) + (-0.125)$
$= -0.25 + (-0.125)$
$= -0.375$

$0.5 + (-0.75) = -0.25.$
$-0.25 + (-0.125) = -0.375.$

53. $x + (-48) + 92 + (-85) + (-31) = -144$
$-72 + (-48) + 92 + (-85) + (-31) = -144$
$[-72 + (-48) + (-85) + (-31)] + 92 = -144$

Replace x with -72.
Group the negative numbers together and the positive numbers together.

$-236 + 92 = -144$
$-144 = -144$

$-72 + (-48) + (-85) + (-31) = -236.$
$-236 + 92 = -144.$

True

$x = -72$ is a solution.

57. $-1020 + 830 = -[|-1020| - 830]$

$= -(1020 - 830)$

$= -190$

-190 ft

Signs are unlike, so subtract their absolute values.

Since -1020 has the largest absolute value the sum is negative.

61. $-2338 + 6313 = 6313 - |-2338|$
$= 6313 - 2338$
$= 3975$

The signs are unlike, so subtract their absolute values.
The sum is positive.

65. $26 + (-6) = 26 - |-6|$
$= 26 - 6$
$= 20$
20%

The signs are unlike, so subtract their absolute values.
The sum is positive.

69. One possible answer: If the signs are alike then the sum will have this same sign. If the signs are unlike then the sum will have the same sign as the number with the largest absolute value.

73. $4(2 + y^3) + 5[2y + (-7)]$ "Increased by" indicates addition.

77. $5^5 = 5 \cdot 5 \cdot 5 \cdot 5 \cdot 5$ 5 is a factor five times.
    $= 3125$ Multiply.

81. $23^5 = 23 \cdot 23 \cdot 23 \cdot 23 \cdot 23$ 23 is a factor five times.

85. $z = 3^3 \cdot 4^2 \cdot 5^2$
    $= 27 \cdot 16 \cdot 25$ Evaluate exponents.
    $= 10800$ Multiply.

EXERCISES 1.4

1. $8 - 4 = 8 + (-4)$ Change to addition.
    $= 4$ Find the difference of their absolute values.

5. $16 - (-4) = 16 + [-(-4)]$ Change to addition.
    $= 16 + 4$ $-(-4) = 4$.
    $= 20$

9. $-5 - (-9) = -5 + [-(-9)]$ Change to addition.
    $= -5 + 9$ $-(-9) = 9$.
    $= 4$ Find the difference of their absolute values.

13. $-15 - (-4) - (-2)$
    $= -15 + [-(-4)] + [-(-2)]$ Change to addition.
    $= -15 + 4 + 2$ $-(-4) = 4; -(-2) = 2$.
    $= -11 + 2$ $-15 + 4 = -11$.
    $= -9$ $-11 + 2 = -9$.

17. $-25 - 48 = -25 + (-48)$ Change to addition.
    $= -73$ Signs are alike, so add their absolute values
    and keep the common sign.

21. $25 - 48 = 25 + (-48)$ Change to addition.
    $= -23$ Find the difference of their absolute values.
    -48 has the larger absolute value so the sum is negative.

25. $-48 - (-25) = -48 + [-(-25)]$ Change to addition.
    $= -48 + 25$ $-(-25) = 25$.
    $= -23$

29. $214 - (385) = 214 + (-385)$ Change to addition.
    $= -171$

33. $-214 - (-385) = -214 + [-(-385)]$ Change to addition.
    $= -214 + 385$ $-(-25) = 25$.
    $= 171$

37. $-12 - 8 + (-6) = -12 + (-8) + (-6)$ Change to addition.
    $= -20 + (-6)$ $-12 + (-8) = -20$.
    $= -26$ $-20 + (-6) = -26$.

Exercises 1.4

41. $-\dfrac{3}{4} - \dfrac{3}{2} = -\dfrac{3}{4} + \left(-\dfrac{3}{2}\right)$ Change to addition.

        $= -\dfrac{3}{4} + \left(-\dfrac{6}{4}\right)$ Least common denominator is 4.

        $= -\dfrac{9}{4}$

45. $-\dfrac{7}{3} - \left(\dfrac{1}{4}\right) = -\dfrac{7}{3} + \left(-\dfrac{1}{4}\right)$ Change to addition.

        $= -\dfrac{28}{12} + \left(\dfrac{-3}{12}\right)$ Least common denominator is 12.

        $= -\dfrac{31}{12}$

49. $-288 - 122 - (-621) + (-111)$
    $= -288 + (-122) + 621 + (-111)$ Change to addition.
    $= [-288 + (-122) + (-111)] + 621$ Group the negative numbers together and group
    $= -521 + 621$ the positive numbers together.
    $= 100$

53. $a = 0.76 - 5.8$
     $= 0.76 + (-5.8)$ Change to addtion.
     $= -5.04$

57. $298.75 - 78.47$
   $= 298.75 + (-78.47)$ Change to addition.
   $= 220.28$

   $\$220.28$

61. $15 - (-9)$ Find the difference of the temperatures.
   $= 15 + 9$ Change to addition.
   $= 24$

   $24°F$

65. $129 - 188$ Find the difference in the number of smoggy days.
   $= 129 + (-188)$ Change to addition.
   $= -59$

   59 days less

69. $7 + (-5 - y)$ "More than" indicates addition; "difference" indicates subtraction.

77. $136 - (14^2 - 10^2)$
    $= 136 - (196 - 100)$ Evaluate exponents.
    $= 136 - 96$ Simplify inside parentheses.
    $= 40$ Subtract.

81. $12^2 - 13(5^2 - 3 \cdot 4 - 8^2) + 12$
    $= 144 - 13(25 - 3 \cdot 4 - 64) + 12$ Evaluate exponents.
    $= 144 - 13(25 - 12 - 64) + 12$ Multiply inside parentheses.
    $= 144 - 13(-51) + 12$ Simplify inside parentheses.
    $= 144 + 663 + 12$ Multiply.
    $= 819$ Add.

EXERCISES 1.5

1. $(-3)(2) = -6$ The product of two unlike signs is negative.

5. $6(-8) = -48$ The product of two unlike signs is negative.

9. $8 \div (-2) = -4$ The signs are unlike, the quotient is negative.

13. $(-14) \div (-2) = 7$ The signs are alike, the quotient is positive.

17. $(-11)(10) = -110$ The product of two unlike signs is negative.

21. $(43)(-6) = -258$ The product of two unlike signs is negative.

25. $125 \div (-25) = -5$ The signs are unlike, the quotient is negative.

29. $(-300) \div (-12) = 25$ The signs are alike, the quotient is positive.

33. $(-4)(-1)^2(-11) = (-4)(1)(-11)$ Evaluate exponent.
                     $= (-4)(-11)$ $(-4)(1) = -4$.
                     $= 44$ $(-4)(-11) = 44$.

37. $(-1)(-3)(-4)(-2) = (3)(-4)(-2)$ $(-1)(-3) = 3$.
                     $= (-12)(-2)$ $(3)(-4) = -12$.
                     $= 24$ $(-12)(-2) = 24$.

41. $\dfrac{-2860}{-10} = 286$ The signs are alike, so the quotient is positive.

45. $z = (-0.8)(0.7)(-4)$
      $= (-0.56)(-4)$ $(-0.8)(0.7) = -0.56$.
      $= 2.24$ $(-0.56)(-4) = 2.24$.

49. $6(-3.5) = -21$ The product of two unlike signs is negative.
    -21 lb

53. $264(-18) = -4752$ The product of two unlike signs is negative.
    -4752 cents $= -\$47.52$

Exercises 1.5

57. $C = \dfrac{5}{9}(F - 32)$

   $= \dfrac{5}{9}(14 - 32)$ Replace F with 14.

   $= \dfrac{5}{9}(-18)$ Subtract.

   $= -10$ Multiply.

   $-10°C$

61. $\dfrac{-26}{5} = -5.2$ The signs are unlike, so the quotient is negative.

65. $\dfrac{11\frac{7}{8} - 17\frac{1}{8} - 21\frac{3}{4} + 5\frac{1}{2} - 6\frac{7}{8}}{5}$ To compute average, add and divide by 5.

   $= \dfrac{11\frac{7}{8} - 17\frac{1}{8} - 21\frac{6}{8} + 5\frac{4}{8} - 6\frac{7}{8}}{5}$ For the fractions in the numerator the least common denominator is 8.

   $= \dfrac{\left(11\frac{7}{8} + 5\frac{4}{8}\right) + \left(-17\frac{1}{8} - 21\frac{6}{8} - 6\frac{7}{8}\right)}{5}$ Group the positive numbers together and group the negative numbers together.

   $= \dfrac{17\frac{3}{8} + \left(-45\frac{6}{8}\right)}{5}$

   $= \dfrac{-28\frac{3}{8}}{5}$

   $= \dfrac{\frac{-227}{8}}{5}$ Change the mixed number to an improper fraction.

   $= \dfrac{-227}{8} \div 5$

   $= \dfrac{-227}{8} \cdot \dfrac{1}{5}$ Invert and multiply.

   $= -\dfrac{227}{40}$

   $= -5\dfrac{27}{40}$

69. $-[(-7)(-y)]$ Product indicates multiplication.

73. $\dfrac{-x}{18}$

Quotient indicates division.

77. $\dfrac{75}{6} = 12.5$

Total weight loss divided by 6.

12.5 lb/month

$\dfrac{75}{26} = 2.9$

Total weight loss divided by 26.

2.9 lb/week

81. One possible answer: Division by zero is undefined.

85. $(2a - b)(3b - 4a)$
 $= [2(-5)-4][3(4) - 4(-5)]$
 $= (-10 - 4)(12 + 20)$
 $= (-14)(32)$
 $= -448$

Replace a with -5 and b with 4.
Multiply inside brackets.
Add or subtract inside parentheses.
Multiply.

89. $y - x$

The second number is replaced with y and the first number is replaced with x.

93. 45 less the square of a number

"Less" indicates subtraction.

97. $45 - x = 36$
 $45 - 9 = 36$
 $36 = 36$
 True

Replace x with 9.
Subtract.

$x = 9$ is a solution.

EXERCISES 1.6

1. $21 - (-8)2 - 5 \cdot 4$
 $= 21 + 16 - 20$
 $= 17$

Multiply.
Add and subtract from left to right.

5. $5^2 - (12 - 7)^2 \cdot 4$
 $= 5^2 - 5^2 \cdot 4$
 $= 25 - 25 \cdot 4$
 $= 25 - 100$
 $= -75$

Subtract inside the parentheses.
Evaluate the exponents.
Multiply.
Subtract.

9. $15a - 3b - 5c$
 $= 15(-7) - 3(-4) - 5(3)$
 $= -105 + 12 - 15$
 $= -108$

Replace a with -7, b with -4 and c with 3.
Multiply from left to right.
Add and subtract from left to right.

13. $5a^2 - 4b^2 - 3c^2$
 $= 5(-7)^2 - 4(-4)^2 - 3(3)^2$
 $= 5(49) - 4(16) - 3(9)$

Replace a with -7, b with -4 and c with 3.
Evaluate the exponents.

Exercises 1.6

$$= 245 - 64 - 27$$
$$= 154$$

Multiply from left to right.
Add and subtract from left to right.

17. $3x - 6 = 24$
   $3(10) - 6 = 24$ Replace x with 10.
   $30 - 6 = 24$ Multiply.
   $24 = 24$ Subtract.
   True

   $x = 10$ is a solution.

21. $-15 - \dfrac{8^2 - 4}{3^2 + 3}$

   $= -15 - \dfrac{64 - 4}{9 + 3}$ Evaluate the exponents.

   $= -15 - \dfrac{60}{12}$ Subtract above the fraction bar and add below the fraction bar.

   $= -15 - 5$ Reduce the fraction.
   $= -20$ Subtract.

25. $\dfrac{1}{2} - \dfrac{\frac{4}{5}}{1 - \frac{1}{3}} \div \dfrac{2}{5}$

   $= \dfrac{1}{2} - \dfrac{\frac{4}{5}}{\frac{3}{3} - \frac{1}{3}} \div \dfrac{2}{5}$ Get a common denominator below the fraction bar.

   $= \dfrac{1}{2} - \dfrac{\frac{4}{5}}{\frac{2}{3}} \div \dfrac{2}{5}$ Subtract below the fraction bar.

   $= \dfrac{1}{2} - \dfrac{4}{5} \div \dfrac{2}{3} \div \dfrac{2}{5}$ Rewrite using a division symbol instead of a fraction bar.

   $= \dfrac{1}{2} - \dfrac{4}{5} \cdot \dfrac{3}{2} \cdot \dfrac{5}{2}$ Divide from left to right by inverting and multiplying.

   $= \dfrac{1}{2} - \dfrac{6}{5} \cdot \dfrac{5}{2}$

   $= \dfrac{1}{2} - 3$

   $= -\dfrac{5}{2}$ Subtract.

29. $\quad -8|125 - 321| - 21^2 + 8(-7)$
$\quad = -8|-196| - 21^2 + 8(-7)$ — Subtract inside the absolute value symbol.
$\quad = -8(196) - 21^2 + 8(-7)$ — $|-196| = 196$.
$\quad = -8(196) - 441 + 8(-7)$ — Evaluate the exponent.
$\quad = -1568 - 441 - 56$ — Multiply from left to right.
$\quad = -2065$ — Subtract from left to right.

33. $\quad 4a - 2b[(4a + 3b) - 3c] - 5a$
$\quad = 4(-5) - 2(-3)\{[4(-5) + 3(-3)] - 3(6)\} - 5(-5)$ — Replace a with -5, b with -3 and c with 6.
$\quad = 4(-5) - 2(-3)(-20 - 9 - 18) - 5(-5)$ — Multiply inside the braces.
$\quad = 4(-5) - 2(-3)(-47) - 5(-5)$ — Subtract inside the parentheses.
$\quad = -20 - 282 + 25$ — Multiply from left to right.
$\quad = -277$ — Add and subtract from left to right.

37. $\quad 2c(a - b)^3 \div c^2$
$\quad = 2(6)[-5 - (-3)]^3 \div 6^2$ — Replace a with -5, b with -3 and c with 6.
$\quad = 2(6)(-2)^3 \div 6^2$ — Subtract inside the brackets.
$\quad = 2(6)(-8) \div 36$ — Evaluate the exponents.
$\quad = -96 \div 36$ — Multiply.
$\quad = -\dfrac{8}{3}$ — Divide.

41. $\quad 8a - 22 = 12a + 17$
$\quad 8\left(-\dfrac{5}{4}\right) - 22 = 12\left(-\dfrac{5}{4}\right) + 17$ — Replace a with $-\dfrac{5}{4}$.
$\quad -10 - 22 = -15 + 17$ — Multiply.
$\quad -32 = 2$ — Subtract and add.
$\quad$ False
$\quad a = -\dfrac{5}{4}$ is not a solution.

45. $\quad 25.6 + \dfrac{18.5 - 20.25}{0.25} + (1.8)(32.7)$

$\quad = 25.6 + \dfrac{-1.75}{0.25} + (1.8)(32.7)$ — Subtract above the fraction bar.

$\quad = 25.6 - 7 + 58.86$ — Divide and multiply from left to right.

$\quad = 77.46$ — Add and subtract from left to right.

49. $\quad 16.4(41.9 - 62.8)^2 + 21.6$
$\quad = 16.4(-20.9)^2 + 21.6$ — Subtract inside the parentheses.
$\quad = 16.4(436.81) + 21.6$ — Evaluate exponent.
$\quad = 7163.684 + 21.6$ — Multiply.
$\quad = 7185.284$ — Add.

53. $\quad 5\dfrac{2}{3} + \dfrac{4\dfrac{1}{2}}{2\dfrac{7}{8} - 3\dfrac{1}{6}} \div 3\dfrac{1}{3}$

Exercises 1.6

$$= 5\frac{2}{3} + \frac{4\frac{1}{2}}{-\frac{7}{24}} \div 3\frac{1}{3}$$

$$= 5\frac{2}{3} + \frac{9}{2} \cdot -\frac{24}{7} \div 3\frac{1}{3}$$

$$= 5\frac{2}{3} - \frac{108}{7} + \frac{10}{3}$$

$$= 5\frac{2}{3} - \frac{108}{7} \cdot \frac{3}{10}$$

$$= 5\frac{2}{3} - \frac{162}{35}$$

$$= \frac{595}{105} - \frac{486}{105}$$

$$= \frac{109}{105}$$

$2\frac{7}{8} - 3\frac{1}{6} = 2\frac{21}{24} - 3\frac{4}{24} = -\frac{7}{24}.$

$4\frac{1}{2} = \frac{9}{2}.$

Multiply.

To divide fractions, invert and multiply.

The common denominator is 105.

Subtract.

57. $a^2 - b^2 - c^2$
$= (-31.2)^2 - (-5.7)^2 - (14.5)^2$
$= 973.44 - 32.49 - 210.25$
$= 730.7$

Replace a with -31.2, b with -5.7 and c with 14.5.
Evaluate the exponents.
Add and subtract from left to right.

61. $\quad 2.3x + 5.6 = 22.298$
$\quad 2.3(7.26) + 5.6 = 22.298$
$\quad 16.698 + 5.6 = 22.298$
$\quad \quad \quad 22.298 = 22.298$
$\quad \quad \quad \quad \quad$ True

$x = 7.26$ is a solution.

Replace x with 7.26.
Multiply.
Add.

65. $C = \frac{5}{9}(F - 32)$

$= \frac{5}{9}(95.9 - 32)$

$= \frac{5}{9}(63.9)$

$= 35.5$

35.5°C

Replace F with 95.9.

Subtract inside the parentheses.

Multiply.

69. $F = \frac{9C + 160}{5}$

$= \frac{91(-20.5) + 160}{5}$

Replace C with -20.5.

$$= \frac{-184.5 + 160}{5}$$ Multiply.

$$= \frac{-24.5}{5}$$ Add.

$$= -4.9$$ Divide.

$-4.9°F$

73. $200(-1.75) + 400(0.80)$
    $= -350 + 320$
    $= -30$ Multiply.
    Add.

Loss of $30.

77. $12x - (y - 8)$ "Decreased by" and "difference" both indicate subtraction.

85. $4^2 - 5[6(4^2 - 8) + 2^3 - (-9)]$
    $= 16 - 5[6(16 - 8) + 8 - (-9)]$ Evaluate the exponents.
    $= 16 - 5[6(8) + 8 - (-9)]$ Subtract inside the parentheses inside the brackets.
    $= 16 - 5(48 + 8 + 9)$ Multiply inside the brackets.
    $= 16 - 5(65)$ Add inside the parentheses.
    $= 16 - 325$ Multiply.
    $= -309$ Subtract.

93. $|13| = 13$ The number is positive, so write the number.

97. $-(-(-25)) = -(25)$ $-(-25) = 25$.
    $= -25$ $-(25) = -25$.

CHAPTER 1 CONCEPT REVIEW

1. True

2. False; 81 is the value of the expression.

3. True

4. False; π is an irrational number approximately equal to 3.14.

5. False; In the expression $6 \div 2 \cdot 3$, division is done before the multiplication.

6. True

7. False; It is the set of all replacements for the variable that satisfy the equation.

8. False; The opposite of -3 is 3.

9. False; "$-x$" should be read "the opposite of x".

10. True

Chapter 1 Test

11. True

12. False; $|0| = 0$ which is neither positive nor negative.

13. False; $-3 + (-4) = -7$.

14. True

15. True

16. False;   If the signs are alike the product is positive and if the signs are unlike the product is negative.

17. False;   The sum of two negative numbers is negative.

18. True

19. True

20. False;   The order of operations for signed numbers is the same as that for whole numbers.

CHAPTER 1 TEST

1. $A = lw$
   $= (2.5)(0.6)$
   $= 1.5$

   Replace l with 2.5 and w with 0.6.
   Multiply.

2. $8x - 4 = 6x + 12$
   $8(8) - 4 = 6(8) + 12$
   $64 - 4 = 48 + 12$
   $60 = 60$
   True

   $x = 8$ is a solution.

   Replace x with 8.
   Multiply.
   Add and subtract.

3. $|3.4| = 3.4$

   The number is positive, so write the number.

4. $-8.56 + (-6.58) + 7$
   $= -15.14 + 7$
   $= -8.14$

   $-8.56 + (-6.58) = -15.14$.
   Add.

5. $2a - b + 15$
   $= 2(27) - 14 + 15$
   $= 54 - 14 + 15$
   $= 55$

   Replace a with 27 and b with 14.
   Multiply.
   Add and subtract from left to right.

6. $91 \div (-13) = -7$

   The signs are unlike, so the quotient is negative.

7. $-(43) = -43$

   The opposite of positive 43 is negative 43.

8. $a^4 = 7^4$
   $ = 2401$

   Replace a with 7.
   Evaluate the exponent.

9. $-75 - 22 - 15 - (-1)$
   $= -97 - 15 - (-1)$
   $= -112 - (-1)$
   $= -111$

   $-75 - 22 = -97$.
   $-97 - 15 = -112$.
   $-112 - (-1) = -112 + 1 = -111$.

10. $-21 + (-84) + 19$
    $= -105 + 19$
    $= -86$

    $-21 + (-84) = -105$.
    Add.

11. $(-27)(-14)(3)$
    $= (378)(3)$
    $= 1134$

    $(-27)(-14) = 378$.
    Multiply.

12. $16 \div 4 \cdot 2 + 4 + 2 \cdot 3$
    $= 4 \cdot 2 + 4 + 2 \cdot 3$
    $= 8 + 4 + 6$
    $= 18$

    Divide.
    Multiply from left to right.
    Add from left to right.

13. $187 - (-122)$
    $= 187 + 122$
    $= 309$

    Change to addition.
    Add.

14. $2a^2 - 5b^3 \div (12c)$
    $= 2(2)^2 - 5(-3)^3 \div [12(-5)]$
    $= 2(4) - 5(-27) \div [12(-5)]$
    $= 8 + 135 \div (-60)$
    $= 8 + (-2.25)$
    $= 5.75$

    Replace a with 2, b with -3 and c with -5.
    Evaluate the exponents.
    Multiply.
    Divide.
    Add.

15. $|-87| = 87$

    The number is negative, so write its opposite.

16. $ 4x + 17 = -9x - 9$
    $ 4(-2) + 17 = -9(-2) - 9$
    $ -8 + 17 = 18 - 9$
    $ 9 = 9$
     True

    $x = -2$ is a solution.

    Replace x with -2.
    Multiply.

17. $5x^2$

    "Product" indicates multiplication.

18. $-(-13.5) = 13.5$

    The opposite of a negative number is a positive number.

19. $15 \div 5 + 12 \cdot 5^2 - 8 \cdot 4$
    $= 15 \div 5 + 12 \cdot 25 - 8 \cdot 4$
    $= 3 + 300 - 32$
    $= 271$

    Evaluate the exponent.
    Multiply and divide from left to right.
    Add and subtract from left to right.

20. $-144 \div (-48) = 3$

    The signs are alike, so the quotient is positive.

21. $\dfrac{n}{4p}$ "Product" indicates multiplication.

22. $(-5.6)(-2)(-4.4)$
    $= (11.2)(-4.4)$ $(-5.6)(-2) = 11.2$.
    $= -49.28$ Multiply.

23. $87 - 40 + 44 - 26$
    $= 47 + 44 - 26$ $87 - 40 = 47$.
    $= 91 - 26$ $47 + 44 = 91$.
    $= 65$ points $91 - 26 = 65$.

24. $60 - (-15)$
    $= 60 + 15$ Change to addition.
    $= 75$

   $75°F$

25. $C = \dfrac{5}{9}(F - 32)$

      $= \dfrac{5}{9}(14 - 32)$ Replace F with 14.

      $= \dfrac{5}{9}(-18)$ Subtract inside the parentheses.

      $= -10$ Multiply.

   $-10°C$

CHAPTER 2

EXERCISES 2.1

1. $13x + 9x = (13 + 9)x$
   $= 22x$

   Distributive property.
   Add.

5. $42a - 48a = (42 - 48)a$
   $= -6a$

   Distributive property.
   Subtract.

9. $3(2x) = (3 \cdot 2)x$
   $= 6x$

   Associative property.
   Multiply.

13. $18x \div 3 = (18 \div 3)x$
    $= 6x$

    Divide the coefficient by the divisor.

17. $-15x - 14x + 5x - 4x$
    $= (-15 - 14 + 5 - 4)x$
    $= -28x$

    Distributive property.
    Add and subtract from left to right.

21. $-74w^2 + 22w^2 - 14w^2 + 84w^2$
    $= (-74 + 22 - 14 + 84)w^2$
    $= 18w^2$

    Distributive property.
    Add and subtract from left to right.

25. $4(2a - 3) = 4(2a) - 4(3)$
    $= 8a - 12$

    Distributive property.
    Multiply.

29. $\dfrac{-148x}{4} = \left(-\dfrac{148}{4}\right)x = -37x$

    Divide the coefficient by the divisor.

33. $8x - a + 4x + 2$
    $= (8 + 4)x - a + 2$
    $= 12x - a + 2$

    Distributive property.
    Add.

37. $17a + 12b - c + 2c - 15b + 7a$
    $= (17 + 7)a + (12 - 15)b + (-1 + 2)c$
    $= 24a - 3b + c$

    Distributive property.
    Add and subtract.

41. $(-484xy) \div (-44) = \left(\dfrac{-484}{-44}\right)xy$
    $= 11xy$

    Divide the coefficient by the divisor.

45. $(-11)(3a - 5) = -11(3a) - 11(-5)$
    $= -33a + 55$

    Distributive property.
    Multiply.

49. $6(2x + 3)$
    $= 6(2x) + 6(3)$
    $= 12x + 18$

    Number of sides times side length.
    Distributive property.
    Multiply.

53. One possible answer: To combine like terms use the distributive property to get the coefficients together and then combine the coefficients.

Exercises 2.2

57. $(4x)(-2) + (-4)(-2x - 5) - 9$
    $= -8x + 8x + 20 - 9$ Multiply.
    $= 11$ $-8x + 8x = 0$.

61. $5^5 = 5 \cdot 5 \cdot 5 \cdot 5 \cdot 5$ 5 is a factor five times.
    $= 3125$ Multiply.

65. $23^5 = 23 \cdot 23 \cdot 23 \cdot 23 \cdot 23$ 23 is a factor five times.

69. $z = 3^3 \cdot 4^2 \cdot 5^2$
    $= 27 \cdot 16 \cdot 25$ Evaluate the exponents.
    $= 10800$ Multiply.

EXERCISES 2.2

1. $a - 3 = 29$
    $a - 3 + 3 = 29 + 3$ Add 3 to both sides.
    $a = 32$ Simplify both sides.

    $\{32\}$

5. $z - 3 = -16$
    $z - 3 + 3 = -16 + 3$ Add 3 to both sides.
    $z = -13$ Simplify both sides.

    $\{-13\}$

9. $y - 9 = -49$
    $y - 9 + 9 = -49 + 9$ Add 9 to both sides.
    $y = -40$ Simplify both sides.

    $\{-40\}$

13. $37 = w + 6$
    $37 - 6 = w + 6 - 6$ Subtract 6 from both sides.
    $31 = w$ Simplify both sides.

    $\{31\}$

17. $l + w = p$
    $l + w - w = p - w$ Subtract w from both sides.
    $l = p - w$ Simplify both sides.

21. $y + 139 = 95$
    $y + 139 - 139 = 95 - 139$ Subtract 139 from both sides.
    $y = -44$ Simplify both sides.

    $\{-44\}$

25. $w + 72 = 0$
    $w + 72 - 72 = 0 - 72$ Subtract 72 from both sides.
    $w = -72$ Simplify both sides.

    $\{-72\}$

29. $\quad x - 341 = 263$
$\quad x - 341 + 341 = 263 + 341 \quad$ Add 341 to both sides.
$\quad\quad\quad\quad\quad\quad x = 604 \quad$ Simplify both sides.
$\{604\}$

33. $\quad\quad -221 = a - 477$
$\quad -221 + 477 = a - 477 + 477 \quad$ Add 477 to both sides.
$\quad\quad\quad\quad 256 = a \quad$ Simplify both sides.
$\{256\}$

37. $\quad x + 34 = 39 - 7$
$\quad x + 34 = 32 \quad$ Simplify right side.
$\quad x + 34 - 34 = 32 - 34 \quad$ Subtract 34 from both sides.
$\quad\quad\quad\quad x = -2 \quad$ Simplify both sides.
$\{-2\}$

41. $\quad x - \dfrac{3}{4} = -\dfrac{3}{2}$

$\quad x - \dfrac{3}{4} + \dfrac{3}{4} = -\dfrac{3}{2} + \dfrac{3}{4} \quad$ Add $\dfrac{3}{4}$ to both sides.

$\quad\quad\quad\quad x = -\dfrac{6}{4} + \dfrac{3}{4} \quad$ 4 is the least common denominator.

$\quad\quad\quad\quad x = -\dfrac{3}{4} \quad$ Add.

$\left\{-\dfrac{3}{4}\right\}$

45. $\quad -\dfrac{4}{5} = c - \dfrac{1}{2}$

$\quad -\dfrac{4}{5} + \dfrac{1}{2} = c - \dfrac{1}{2} + \dfrac{1}{2} \quad$ Add $\dfrac{1}{2}$ to both sides.

$\quad -\dfrac{8}{10} + \dfrac{5}{10} = c \quad$ 10 is the least common denominator.

$\quad -\dfrac{3}{10} = c \quad$ Add.

$\left\{-\dfrac{3}{10}\right\}$

49. $\quad\quad z + 17.86 = 5.49$
$\quad z + 17.86 - 17.86 = 5.49 - 17.86 \quad$ Subtract 17.86 from both sides.
$\quad\quad\quad\quad z = -12.37 \quad$ Simplify both sides.
$\{-12.37\}$

53. $\quad\quad\quad c - 42.8 = -50.3$
$\quad c - 42.8 + 42.8 + 42.8 = -50.3 + 42.8 \quad$ Add 42.8 to both sides.
$\quad\quad\quad\quad\quad c = -7.5 \quad$ Simplify both sides.
$\{-7.5\}$

Exercises 2.3

57. $t - m = 5$
     $t - m + m = 5 + m$ Add m to both sides.
     $t = 5 + m$ Simplify left side.

61. number: x

     $x - 33 = 18$ "less than" indicates subtraction.
     $x - 33 + 33 = 18 + 33$ Add 33 to both sides.
     $x = 51$ Simplify both sides.

     The number is 51.

65. Cost + markup = selling price   Formula.
     $C + 83.13 = 498.79$ Replace markup with 83.13 and selling price with 498.79.
     $C + 83.13 - 83.13 = 498.79 - 83.13$ Subtract 83.13 from both sides.
     $C = 415.66$ Simplify both sides.
     $415.66

69. groceries: C

     $19 + C = 43$
     $-19 + 19 + C = -19 + 43$ Add -19 to both sides.
     $C = 24$ Simplify both sides.
     $24

73. One possible answer: Subtracting 2.5 is the same as adding the opposite of 2.5 which is -2.5.

77. $22 = 17 + 5 - 15b + 12b + 2b$
     $22 = 22 - b$ Simplify right side.
     $-22 + 22 = -22 + 22 - b$ Add -22 to both sides.
     $0 = b$ Simplify both sides.
     $\{0\}$

81. $-25 + 47 = |47| - |-25|$ The signs are unlike so take the difference of their absolute values.
     $= 47 - 25$
     $= 22$ 47 has the larger absolute value so the result is positive.

85. $z = -65 - 45 + (-32)$
     $= -65 + (-45) + (-32)$ Change subtraction to addition.
     $= -142$ Add.

EXERCISES 2.3

1. $2x = 50$

    $\dfrac{2x}{2} = \dfrac{50}{2}$ Divide both sides by 2.

    $x = 25$ Simplify both sides.

    $\{25\}$

5. $\quad \dfrac{x}{2} = 50$

   $2\left(\dfrac{x}{2}\right) = 2(50)$ \qquad Multiply both sides by 2.

   $x = 100$ \qquad Simplify both sides.

   $\{100\}$

9. $\quad \dfrac{y}{6} = 11$

   $6\left(\dfrac{y}{6}\right) = 6(11)$ \qquad Multiply both sides by 6.

   $y = 66$ \qquad Simplify both sides.

   $\{66\}$

13. $\quad \dfrac{-15w}{-8} = 30$

    $\dfrac{8}{15}\left(\dfrac{-15w}{-8}\right) = \dfrac{8}{15}(30)$ \qquad Multiply both sides by $\dfrac{8}{15}$.

    $w = 16$ \qquad Simplify both sides.

    $\{16\}$

17. $\quad \dfrac{t}{22} = 88$

    $22\left(\dfrac{t}{22}\right) = 22(88)$ \qquad Multiply both sides by 22.

    $t = 1936$ \qquad Simplify both sides.

    $\{1936\}$

21. $\quad \dfrac{w}{-6} = 30$

    $-6\left(\dfrac{w}{-6}\right) = -6(30)$ \qquad Multiply both sides by -6.

    $w = -180$ \qquad Simplify both sides.

    $\{-180\}$

25. $\quad \dfrac{y}{16} = 27$

    $16\left(\dfrac{y}{16}\right) = 16(27)$ \qquad Multiply both sides by 16.

    $y = 432$ \qquad Simplify both sides.

    $\{432\}$

Exercises 2.3

29. $\dfrac{z}{11} = 50$

$11\left(\dfrac{z}{11}\right) = 11(50)$ Multiply both sides by 11.

$z = 550$ Simplify both sides.

$\{550\}$

33. $5x = m$

$\dfrac{5x}{5} = \dfrac{m}{5}$ Divide both sides by 5.

$x = \dfrac{m}{5}$ Simplify both sides.

37. $7w = -9.03$

$\dfrac{7w}{7} = \dfrac{-9.03}{7}$ Divide both sides by 7.

$w = -1.29$ Simplify both sides.

$\{-1.29\}$

41. $6x = -4$

$\dfrac{6x}{6} = \dfrac{-4}{6}$ Divide both sides by 6.

$x = -\dfrac{2}{3}$ Simplify both sides.

45. $\dfrac{t}{3} = -\dfrac{1}{5}$

$3\left(\dfrac{t}{3}\right) = 3\left(-\dfrac{1}{5}\right)$ Multiply both sides by 3.

$t = -\dfrac{3}{5}$ Simplify both sides.

$\left\{-\dfrac{3}{5}\right\}$

49. $\dfrac{3}{5}t = -\dfrac{7}{10}$

$\dfrac{5}{3}\left(\dfrac{3}{5}t\right) = \dfrac{5}{3}\left(-\dfrac{7}{10}\right)$ Multiply both sides by $\dfrac{5}{3}$.

$t = -\dfrac{7}{6}$ Simplify both sides.

$\left\{-\dfrac{7}{6}\right\}$

53. $\dfrac{A}{w} = l$

$w\left(\dfrac{A}{w}\right) = wl$ Multiply both sides by w.

$A = wl$ Simplify the left side.

57. number: x

$\dfrac{x}{53} = 73$

$53\left(\dfrac{x}{53}\right) = 53(73)$ Multiply both sides by 53.

$x = 3869$ Simplify both sides.

The number is 3869.

61. number: x

$\dfrac{x}{49} = 63$

$49\left(\dfrac{x}{49}\right) = 49(63)$ Multiply both sides by 49.

$x = 3087$ Simplify both sides.

The number is 3087.

65. width: w

$A = \text{width} \cdot \text{length}$ Formula.
$6765 = w \cdot 123$ Replace A with 6765 and length with 123.

$\dfrac{6765}{123} = \dfrac{w \cdot 123}{123}$ Divide both sides by 123.

$55 = w$ Simplify both sides.

The width is 55 m.

69. number of togas: t

$5.2t = 117$

$\dfrac{5.2t}{5.2} = \dfrac{117}{5.2}$ Divide both sides by 5.2.

$t = 22.5$ Simplify both sides.

22.5 togas

73. $\dfrac{3}{4}x - \dfrac{1}{2} - \dfrac{3}{5}x - \dfrac{3}{5} = -\dfrac{7}{8}$

$40\left(\dfrac{3}{5}x - \dfrac{1}{2} - \dfrac{3}{5}x - \dfrac{3}{5}\right) = 40\left(-\dfrac{7}{8}\right)$ Multiply both sides by 40.

$$40\left(\frac{3}{4}x\right) - 40\left(\frac{1}{2}\right) - 40\left(\frac{3}{5}x\right) - 40\left(\frac{3}{5}\right) = 40\left(-\frac{7}{8}\right)$$ Distribute.

$$30x - 20 - 24x - 24 = -35$$ Multiply.
$$6x - 44 = -35$$ Combine like terms on the left side.
$$6x - 44 + 44 = -35 + 44$$ Add 44 to both sides.
$$6x = 9$$ Simplify both sides.

$$\frac{6x}{6} = \frac{9}{6}$$ Divide both sides by 6.

$$x = \frac{3}{2}$$ Simplify both sides.

$$\left\{\frac{3}{2}\right\}$$

81. $-(-67) - (-89) + 64 - (-43) - 62$
 $= 67 + 89 + 64 + 43 + (-62)$ Change the subtractions to additions.
 $= 201$ Add from left to right.

85. $c - b = -48 - 17$ Replace c with -48 and b with 17.
 $= -65$ Subtract.

EXERCISES 2.4

1. $$7x + 5 = 40$$
$$7x + 5 - 5 = 40 - 5$$ Subtract 5 from both sides.
$$7x = 35$$ Simplify.

$$\frac{7x}{7} = \frac{35}{7}$$ Divide both sides by 7.

$$x = 5$$ Simplify.
$$\{5\}$$

5. $$2y + 17 = -8$$
$$2y + 17 - 17 = -8 - 17$$ Subtract 17 from both sides.
$$2y = -25$$ Simplify.

$$\frac{2y}{2} = \frac{-25}{2}$$ Divide both sides by 2.

$$y = -\frac{25}{2}$$ Simplify.

$$\left\{-\frac{25}{2}\right\}$$

9. $$0 = -3x + 15$$
$$0 - 15 = -3x + 15 - 15$$ Subtract 15 from both sides.
$$-15 = -3x$$ Simplify.

$$\frac{-15}{-3} = \frac{-3x}{-3}$$ Divide both sides by -3.

$$5 = x$$ Simplify.
$$\{5\}$$

13.
$$ax + b = c$$
$$3x + 4 = 1$$ Replace a with 3, b with 4 and c with 1.
$$3x + 4 - 4 = 1 - 4$$ Subtract 4 from both sides.
$$3x = -3$$ Simplify.
$$\frac{3x}{3} = \frac{-3}{3}$$ Divide both sides by 3.
$$x = -1$$ Simplify.
$\{-1\}$

17.
$$29y + 2 = 205$$
$$29y + 2 - 2 = 205 - 2$$ Subtract 2 from both sides.
$$29y = 203$$ Simplify.
$$\frac{29y}{29} = \frac{203}{29}$$ Divide both sides by 29.
$$y = 7$$ Simplify.
$\{7\}$

21.
$$67w + 34 = 369$$
$$67w + 34 - 34 = 369 - 34$$ Subtract 34 from both sides.
$$67w = 335$$ Simplify.
$$\frac{67w}{67} = \frac{335}{67}$$ Divide both sides by 67.
$$w = 5$$ Simplify.
$\{5\}$

25.
$$4 = 22x - 150$$
$$4 + 150 = 22x - 150 + 150$$ Add 150 to both sides.
$$154 = 22x$$ Simplify.
$$\frac{154}{22} = \frac{22x}{22}$$ Divide both sides by 22.
$$7 = x$$ Simplify.
$\{7\}$

29.
$$7y - 4 = 4y + 20$$
$$7y - 4 + 4 = 4y + 20 + 4$$ Add 4 to both sides.
$$7y = 4y + 24$$ Simplify.
$$-4y + 7y = -4y + 4y + 24$$ Add $-4y$ to both sides.
$$3y = 24$$ Simplify.
$$\frac{3y}{3} = \frac{24}{3}$$ Divide both sides by 3.
$$y = 8$$ Simplify.
$\{8\}$

33.
$$21 - 2t = 8t - 19$$
$$21 - 2t + 2t = 2t + 8t - 19$$ Add $2t$ to both sides.
$$21 = 10t - 19$$ Simplify.
$$21 + 19 = 10t - 19 + 19$$ Add 19 to both sides.

Exercises 2.4

$$40 = 10t$$ Simplify.

$$\frac{40}{10} = \frac{10t}{10}$$ Divide both sides by 10.

$$4 = t$$ Simplify.

{4}

37.
$$19 - 2x - 5 = 5x + 7$$
$$-2x + 14 = 5x + 7$$ Combine like terms.
$$-2x + 14 - 14 = 5x + 7 - 14$$ Subtract 14 from both sides.
$$-2x = 5x - 7$$ Simplify.
$$-5x - 2x = -5x + 5x - 7$$ Add $-5x$ to both sides.
$$-7x = -7$$ Simplify.

$$\frac{-7x}{-7} = \frac{-7}{-7}$$ Divide both sides by -7.

$$x = 1$$ Simplify.

{1}

41.
$$39x + 32 = 31x$$
$$-39x + 39x + 32 = -39x + 31x$$ Add $-39x$ to both sides.
$$32 = -8x$$ Simplify.

$$\frac{32}{-8} = \frac{-8x}{-8}$$ Divide both sides by -8.

$$-4 = x$$ Simplify.

{-4}

45.
$$5x + 7x = 4x$$
$$12x = 4x$$ Combine like terms.
$$12x - 4x = 4x - 4x$$ Subtract $4x$ from both sides.
$$8x = 0$$ Simplify.

$$\frac{8x}{8} = \frac{0}{8}$$ Divide both sides by 8.

$$x = 0$$ Simplify.

{0}

49.
$$3x + 15 + 21 = 22 + 4x$$
$$3x + 36 = 22 + 4x$$ Combine like terms.
$$-3x + 3x + 36 = 22 - 3x + 4x$$ Subtract $3x$ from both sides.
$$36 = 22 + x$$ Simplify.
$$36 - 22 = 22 - 22 + x$$ Subtract 22 from both sides.
$$14 = x$$ Simplify.

{14}

53.
$$37q + 9 - 8q = 8q - 9 - 87$$
$$29q + 9 = 8q - 96$$ Combine like terms.
$$29q + 9 - 9 = 8q - 96 - 9$$ Subtract 9 from both sides.
$$29q = 8q - 105$$ Simplify.
$$-8q + 29q = -8q + 8q - 105$$ Add $-8q$ to both sides.

$21q = -105$ — Simplify.

$\dfrac{21q}{21} = \dfrac{-105}{21}$ — Divide both sides by 21.

$q = -5$ — Simplify.

$\{-5\}$

57.
$2d = t^2 a + 2tv$ — Formula.
$2(240) = 3^2 a + 2(3)(20)$ — Replace d with 240, v with 20 and t with 3.
$480 = 9a + 120$ — Simplify.
$480 - 120 = 9a + 120 - 120$ — Subtract 120 from both sides.
$360 = 9a$ — Simplify.

$\dfrac{360}{9} = \dfrac{9a}{9}$ — Divide both sides by 9.

$40 = a$ — Simplify.

61. number: x

$15x - 87 = -39$ — "Difference" indicates subtraction.
$15x - 87 + 87 = -39 + 87$ — Add 87 to both sides.
$15x = 48$ — Simplify.

$\dfrac{15x}{15} = \dfrac{48}{15}$ — Divide both sides by 15.

$x = 3.2$ — Simplify.

The number is 3.2.

65. hours: h

$22h + 125 = 301$ — Labor + Parts = Total bill.
$22h + 125 - 125 = 301 - 125$ — Subtract 125 from both sides.
$22h = 176$ — Simplify.

$\dfrac{22h}{22} = \dfrac{176}{22}$ — Divide both sides by 22.

$h = 8$ — Simplify.

8 hrs

69. One possible answer: If you combine like terms first, there will be less terms to move using the properties of equality.

73.
$A = \dfrac{1}{2} h(a + b)$

$2A = 2\left[\dfrac{1}{2} h(a + b)\right]$ — Multiply both sides by 2.

$2A = h(a + b)$ — Simplify.

$\dfrac{2A}{h} = \dfrac{h(a + b)}{h}$ — Divide both sides by h.

Exercises 2.5

$$\frac{2A}{h} = a + b \qquad \text{Simplify.}$$

$$\frac{2A}{h} - a = a - a + b \qquad \text{Subtract } a \text{ from both sides.}$$

$$\frac{2A}{h} - a = b \qquad \text{Simplify.}$$

77. $(-2.3)(-0.3)(5.6)$
 $= (0.69)(5.6)$ $\qquad (-2.3)(-0.3) = 0.69$
 $= 3.864$ Multiply.

81. $c \div a = -36 \div (-6)$ Replace c with -36 and a with -6.
 $= 6$ The signs are alike, so the quotient is positive.

EXERCISES 2.5

1. $RB = A$ Formula.
 $R(30) = 15$ Replace B with 30 and A with 15.

 $$\frac{R(30)}{30} = \frac{15}{30} \qquad \text{Divide both sides by 30.}$$

 $R = 0.5 = 50\%$ Simplify.

5. $RB = A$ Formula.
 $(0.50)B = 13$ Replace R with $50\% = 0.50$ and A with 13.

 $$\frac{0.50B}{0.50} = \frac{13}{0.50} \qquad \text{Divide both sides by 0.50.}$$

 $B = 26$ Simplify.

9. $RB = A$ Formula.
 $(0.30)B = 48$ Replace R with $30\% = 0.30$ and A with 48.

 $$\frac{0.30B}{0.30} = \frac{48}{0.30} \qquad \text{Divide both sides by 0.30.}$$

 $B = 160$ Simplify.

13. $RB = A$ Formula.
 $(1.50)(30) = A$ Replace R with $150\% = 1.50$ and B with 30.
 $45 = A$ Multiply.

17. $RB = A$ Formula.
 $(0.39)(60) = A$ Replace R with $39\% = 0.39$ and B with 60.
 $23.4 = A$ Multiply.

21. $RB = A$ Formula.
 $R(75) = 25$ Replace B with 75 and A with 25.

 $$\frac{R(75)}{75} = \frac{25}{75} \qquad \text{Divide both sides by 75.}$$

 $R = 0.33\frac{1}{3} = 33\frac{1}{3}\%$ Simplify.

25. $\quad RB = A$ Formula.
$\quad\;\; R(200) = 18.6$ Replace B with 200 and A with 18.6.

$\quad\;\; \dfrac{R(200)}{200} = \dfrac{18.6}{200}$ Divide both sides by 200.

$\quad\;\; R = 0.093 = 9.3\%$ Simplify.

29. $\quad RB = A$ Formula.
$\quad\;\; R(360) = 43.2$ Replace B with 360 and A with 43.2.

$\quad\;\; \dfrac{R(360)}{360} = \dfrac{43.2}{360}$ Divide both sides by 360.

$\quad\;\; R = 0.12 = 12\%$ Simplify.

33. $\quad RB = A$ Formula.
$\quad\;\; 0.0075B = 21$ Replace R with $\frac{3}{4}\% = 0.0075$ and A with 21.

$\quad\;\; \dfrac{0.0075B}{0.0075} = \dfrac{21}{0.0075}$ Divide both sides by 0.0075.

$\quad\;\; B = 2800$ Simplify.

37. $\quad RB = A$ Formula.
$\quad\;\; 0.43B = 39$ Replace R with $43\% = 0.43$ and A with 39.

$\quad\;\; \dfrac{0.43B}{0.43} = \dfrac{39}{0.43}$ Divide both sides by 0.43.

$\quad\;\; B = 90.7$ Round answer to nearest tenth.

41. $\quad RB = A$ Formula.
$\quad\;\; (0.093)(940) = A$ Replace R with $9.3\% = 0.093$ and B with 940.
$\quad\;\; 87.4 = A$ Multiply and round answer to nearest tenth.

45. $\quad RB = A$ Formula.
$\quad\;\; (0.623)(18) = A$ Replace R with $62.3\% = 0.623$ and B with 18.
$\quad\;\; 11.2 = A$ Multiply and round answer to nearest tenth.

49. $\quad RB = A$ Formula.
$\quad\;\; (1.213)(73) = A$ Replace R with $121.3\% = 1.213$ and B with 73.
$\quad\;\; 88.5 = A$ Multiply and round answer to nearest tenth.

53. $\quad RB = A$ Formula.
$\quad\;\; 0.22B = 73.5$ Replace R with $22\% = 0.22$ and A with 73.5.

$\quad\;\; \dfrac{0.22B}{0.22} = \dfrac{73.5}{0.22}$ Divide both sides by 0.22.

$\quad\;\; B = 334.09$ Simplify and round answer to nearest hundredth.

57. $\quad RB = A$ Formula.
$\quad\;\; R(24) = 16$ Replace B with 24 and A with 16.

Exercises 2.5

$$\frac{R(24)}{24} = \frac{16}{24}$$ Divide both sides by 24.

$$R = 0.66\frac{2}{3} = 66\frac{2}{3}\%$$ Simplify.

61. $RB = A$ Formula.
     $0.24B = 275$ Replace R with $24\% = 0.24$ and A with 275.

$$\frac{0.24B}{0.24} = \frac{275}{0.24}$$ Divide both sides by 0.24.

     $B = 1145.83$ Simplify.

$1145.83

65. $RB = A$ Formula.
     $1.15B = 362.25$ Cost + 15% = 1.15%.

$$\frac{1.15B}{1.15} = \frac{362.25}{1.15}$$ Divide both sides by 1.15.

     $B = 315$ Simplify.

$315

69. $RB = A$ Formula.
     $(1.15)(520) = A$ Cost + 15% = 1.15.
     $598 = A$ Multiply.

$598

73. $RB = A$ Formula.
     $0.15B = 11$ Replace R with $15\% = 0.15$ and A with 11.

$$\frac{0.15B}{0.15} = \frac{11}{0.15}$$ Divide both sides by 0.15.

$$B = 73\frac{1}{3}$$ Simplify.

$73\frac{1}{3}$ tons

77. One possible answer: The rate (Y) is the number before the word "percent", the base (Z) is the number that follows the word "of" and the amount (X) is the number compared with the base.

81. $RB = A$ Formula.
     $0.85(0.75B) = 545$ 100% − 25% = 75% = 0.75 and 100% − 15% = 85% = 0.85.

     $0.6375B = 545$ Multiply.

$$\frac{0.6375B}{0.6375} = \frac{545}{0.6375}$$ Divide both sides by 0.6375.

     $B = 854.90$ Simplify.

$854.90

85. $(-4)(-64) = 256$ — The signs are alike, so the product is positive.

89. $(-900) \div 36 = -25$ — The signs are unlike, so the quotient is negative.

EXERCISES 2.6

1. $x - 7 > 13$
   $x - 7 + 7 > 13 + 7$ — Add 7 to both sides.
   $x > 20$ — Simplify.

   $\{x \mid x > 20\}$

5. $5x \leq -45$

   $\dfrac{5x}{5} \leq \dfrac{-45}{5}$ — Divide both sides by 5. Since 5 is positive, the inequality does not reverse.

   $x \leq -9$ — Simplify.

   $\{x \mid x \leq -9\}$

9. $-6x < 54$

   $\dfrac{-6x}{-6} > \dfrac{54}{-6}$ — Divide both sides by -6. We must reverse the inequality sign when dividing by a negative number.

   $x > -9$ — Simplify.

   $\{x \mid x > -9\}$

13. $x + 8 \geq 16$
    $x + 8 - 8 \geq 16 - 8$ — Subtract 8 from both sides.
    $x \geq 8$ — Simplify.

    $\{x \mid x \geq 8,\ x \in J\}$

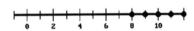

17. $4x + 1 \geq 5$
    $4x + 1 - 1 \geq 5 - 1$ — Subtract 1 from both sides.
    $4x \geq 4$ — Simplify.

    $\dfrac{4x}{4} \geq \dfrac{4}{4}$ — Divide both sides by 4. Since 4 is positive, the inequality does not reverse.

    $x \geq 1$ — Simplify.

    $\{x \mid x \geq 1\}$

21. $9x + 6 > -12$
    $9x + 6 - 6 > -12 - 6$ — Subtract 6 from both sides.
    $9x > -18$ — Simplify.

Exercises 2.6

$$\frac{9x}{9} > \frac{-18}{9}$$ Divide both sides by 9. Since 9 is positive, the inequality does not reverse.

$$x > -2$$ Simplify.

$$\{x \mid x > -2\}$$

25. $$-10x + 8 > 48$$
$$-10x + 8 - 8 > 48 - 8$$ Subtract 8 from both sides.
$$-10x > 40$$ Simplify.

$$\frac{-10x}{-10} < \frac{40}{-10}$$ Divide both sides by -10. Since -10 is negative, the inequality reverses.

$$x < -4$$ Simplify.
$$\{x \mid x < -4\}$$

29. $$9x - 5 > -23$$
$$9x - 5 + 5 > -23 + 5$$ Add 5 to both sides.
$$9x > -18$$ Simplify.

$$\frac{9x}{9} > \frac{-18}{9}$$ Divide both sides by 9. Since 9 is positive, the inequality does not reverse.

$$x > -2$$ Simplify.
$$\{x \mid x > -2\}$$

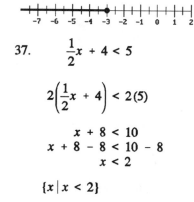

33. $$-8x + 2 \geq 26$$
$$-8x + 2 - 2 \geq 26 - 2$$ Subtract 2 from both sides.
$$-8x \geq 24$$ Simplify.

$$\frac{-8x}{-8} \leq \frac{24}{-8}$$ Divide both sides by -8. Since -8 is negative, the inequality reverses.

$$x \leq -3$$ Simplify.

$$\{x \mid x \leq -3\}$$

37. $$\frac{1}{2}x + 4 < 5$$

$$2\left(\frac{1}{2}x + 4\right) < 2(5)$$ Multiply both sides by 2 to clear the fractions.

$$x + 8 < 10$$ Multiply using the distributive property.
$$x + 8 - 8 < 10 - 8$$ Subtract 8 from both sides.
$$x < 2$$ Simplify.

$$\{x \mid x < 2\}$$

41. $$-3x + 2 \geq 5$$
$$-3x + 2 - 2 \geq 5 - 2$$ Subtract 2 from both sides.
$$-3x \geq 3$$ Simplify.

$$\frac{-3x}{-3} \leq \frac{3}{-3}$$

Divide both sides by −3. Since −3 is negative, the inequality reverses.

$$x \leq -1$$

Simplify.

$$\{x \mid x \leq -1\}$$

45. $\dfrac{x}{-2} + 3 < 4$

$-2\left(\dfrac{x}{-2} + 3\right) > -2(4)$

Multiply both sides by −2 to clear the fractions. Since −2 is negative, the inequality reverses.

$x - 6 > -8$

Multiply using the distributive property.

$x - 6 + 6 > -8 + 6$

Add 6 to both sides.

$x > -2$

Simplify.

$\{x \mid x > -2\}$

49. $5x - 8 \leq 10$

$5x - 8 + 8 \leq 10 + 8$

Add 8 to both sides.

$5x \leq 18$

Simplify.

$\dfrac{5x}{5} \leq \dfrac{18}{5}$

Divide both sides by 5. Since 5 is positive, the inequality does not reverse.

$x \leq \dfrac{18}{5}$

Simplify.

$\left\{x \mid x \leq \dfrac{18}{5}\right\}$

53. $\dfrac{2}{3}x + 4 < 2$

$3\left(\dfrac{2}{3}x + 4\right) < 3(2)$

Multiply both sides by 3 to clear the fractions.

$2x + 12 < 6$

Multiply using the distributive property.

$2x + 12 - 12 < 6 - 12$

Subtract 12 from both sides.

$2x < -6$

Simplify.

$\dfrac{2x}{2} < \dfrac{-6}{2}$

Divide both sides by 2. Since 2 is positive, the inequality does not reverse.

$x < -3$

Simplify.

$\{x \mid x < -3\}$

```
+--+--+--+--o--+--+--+--+--+--+
-7 -6 -5 -4 -3 -2 -1  0  1  2
```

57. number of bags: x

$113x + 175 \leq 3000$

Cement weight + operator's weight = total weight.

$113x + 175 - 175 \leq 3000 - 175$

Subtract 175 from both sides.

$113x \leq 2825$

Simplify.

Exercises 2.6

$$\frac{113x}{113} \leq \frac{2825}{113}$$ Divide both sides by 113.

$$x \leq 25$$ Simplify.

At most 25 bags.

61. number: x

$$5x - 6 < 34$$
$$5x - 6 + 6 < 34 + 6$$
$$5x < 40$$

"Difference" indicates subtraction.
Add 6 to both sides.
Simplify.

$$\frac{5x}{5} < \frac{40}{5}$$ Divide both sides by 5.

$$x < 8$$ Simplify.

Number less than 8.

65. width: x

$$A \leq 144$$
$$18w \leq 144$$ Replace length with 18.

$$\frac{18w}{18} \leq \frac{144}{18}$$ Divide both sides by 18.

$$w \leq 8$$ Simplify.

$$0 < w \leq 8 \text{ in}$$

69. days attended: x

$$2x + 486 \geq 540$$
$$2x + 486 - 486 \geq 540 - 486$$
$$2x \geq 54$$

Attendance points + test points = total points.
Subtract 486 from both sides.
Simplify.

$$\frac{2x}{2} \geq \frac{54}{2}$$ Divide both sides by 2.

$$x \geq 27$$ Simplify.

At least 27 days.

73. One possible answer: First multiply by 3 to clear the fractions. Use the distributive property to multiply. The inequality would now look like $6 - y \leq 15$. Next subtract 6 from both sides to isolate the variable. Last divide both sides by -1 to get a coefficient of 1 for the variable. This division would cause the inequality to reverse since -1 is a negative number.

$$6 - y \leq 15$$
$$-6 + 6 - y \leq 15 - 6$$
$$-y \leq 9$$
$$\frac{-y}{-1} \geq \frac{9}{-1}$$
$$y \geq -9$$

$\{y \mid y \geq -9\}$

77. Ms. Smythe's age: $x - 13$
    Mr. Moss' age: x
    Mr. Young's age: $\frac{1}{2}x$

$(x - 13) + x \geq 50 + \frac{1}{2}x$ On the left add Ms. Smythe's and Mr. Moss' age; on the right add 50 to Mrs. Young's age.

$2x - 13 \geq 50 + \frac{1}{2}x$ Combine like terms.

$2(2x - 13) \geq 2\left(50 + \frac{1}{2}x\right)$ Multiply by 2 to clear the fractions.

$4x - 26 \geq 100 + x$ Multiply using the distributive property.
$4x - 26 \geq 100 + 26 + x$ Add 26 to both sides.
$4x \geq 126 + x$ Simplify.
$4x - x \geq 126 + x - x$ Subtract x from both sides.
$3x \geq 126$ Simplify.

$\frac{3x}{3} \geq \frac{126}{3}$ Divide both sides by 3.

$x \geq 42$ Simplify.

Mr. Moss is at least 42 years old.

81. $\dfrac{2}{5} - \dfrac{1 - \frac{7}{8}}{\frac{1}{3} - \frac{3}{5}} - \dfrac{1}{4}$

$= \dfrac{2}{5} - \dfrac{\frac{8}{8} - \frac{7}{8}}{\frac{5}{15} - \frac{9}{15}} - \dfrac{1}{4}$ $1 = \frac{8}{8}; \frac{1}{3} = \frac{5}{15}; \frac{3}{5} = \frac{9}{15}.$

$= \dfrac{2}{5} - \dfrac{\frac{1}{8}}{-\frac{4}{15}} - \dfrac{1}{4}$ $\frac{8}{8} - \frac{7}{8} = \frac{1}{8}; \frac{5}{15} - \frac{9}{15} = -\frac{4}{15}.$

$= \dfrac{2}{5} - \dfrac{1}{8} \cdot \dfrac{-15}{4} - \dfrac{1}{4}$ Invert and multiply.

$= \dfrac{2}{5} + \dfrac{15}{32} - \dfrac{1}{4}$ $-\dfrac{1}{8} \cdot \dfrac{-15}{4} = \dfrac{15}{32}.$

Chapter 2 Concept Review

$= \dfrac{64}{160} + \dfrac{75}{160} - \dfrac{40}{160}$ 160 is the least common denominator.

$= \dfrac{99}{160}$ Add.

85. $\dfrac{a}{b}(c) = \dfrac{-15}{-24}\left(-\dfrac{2}{3}\right)$ Replace a with -15, b with -24 and c with $-\dfrac{2}{3}$.

$= \dfrac{-5 \cdot 1}{12 \cdot 1}$ Reduce -15 and 3 by a common factor of 3; reduce -24 and -2 by a common factor of -2.

$= -\dfrac{1}{12}$ Multiply.

CHAPTER 2 CONCEPT REVIEW

1. True

2. True

3. False;
$$6x = 4x$$
$$6x - 4x = 0$$
$$2x = 0$$
$$\dfrac{2x}{2} = \dfrac{0}{2}$$
$$x = 0$$
$$\{0\}$$

4. True

5. False;   It is possible to have none, one or more than one solution.

6. True

7. True

8. True

9. False;   There is no solution since division by 0 is undefined.

10. False; $100 + 0.10(100) = 110$
$110 - 0.10(110) = 110 - 11 = 99$
He is now making $99.

11. False;   To decrease by 110% is to decrease by more than the whole.

12. False; $(0.074)(400) = 29.6$

13. True

14. True

15. True

16. False; $-2x + 5 > 13$
    $-2x > 8$
    $x < -4$

    The largest integer in the solution set is -5.

17. False; $-2x + 5 > 13$
    $-2x > 8$
    $x < -4$

    There is no smallest integer in the solution set.

18. False; The same addition and subtraction properties of equality are used, but the multiplication and division properties are different.

19. True

20. True

CHAPTER 2 TEST

1. $RB = A$ Formula.
   $(0.23)(17) = A$ Replace R with $23\% = 0.23$ and B with 17.
   $3.91 = A$ Multiply.

2. $\dfrac{s}{9} = -\dfrac{7}{18}$

   $9\left(\dfrac{s}{9}\right) = 9\left(-\dfrac{7}{18}\right)$ Multiply both sides by 9.

   $s = -\dfrac{7}{2}$ Simplify.

   $\left\{-\dfrac{7}{2}\right\}$

3. $13 - z = 28$
   $-13 + 13 - z = -13 + 28$ Add -13 to both sides.
   $-z = 15$ Simplify.

   $\dfrac{-z}{-1} = \dfrac{15}{-1}$ Divide both sides by -1.

   $z = -15$ Simplify.

   $\{-15\}$

4. $b + ax = c$
   $24 + 10x = -15$ Replace a with 10, b with 24 and c with -15.
   $-24 + 24 + 10x = -24 - 15$ Add -24 to both sides.
   $10x = -39$ Simplify.

$$\frac{10x}{10} = \frac{-39}{10}$$ Divide both sides by 10.

$$x = -3.9$$ Simplify.

$\{-3.9\}$

5. $\quad x + 66 = 34$
$\quad x + 66 - 66 = 34 - 66$ Subtract 66 from both sides.
$\quad\quad\quad\quad x = -32$ Simplify.

$\{-32\}$

6. $\quad 0.2t + 5.8 = 1.8$
$\quad 0.2t + 5.8 - 5.8 = 1.8 - 5.8$ Subtract 5.8 from both sides.
$\quad\quad\quad\quad 0.2t = -4$ Simplify.

$$\frac{0.2t}{0.2} = \frac{-4}{0.2}$$ Divide both sides by 0.2.

$\quad\quad\quad\quad t = -20$ Simplify.

$\{-20\}$

7. $\quad a + 8.5 = 2.3$
$\quad a + 8.5 - 8.5 = 2.3 - 8.5$ Subtract 8.5 from both sides.
$\quad\quad\quad\quad a = -6.2$ Simplify.

$\{-6.2\}$

8. $x > 37, \; x \in J$ x is an integer.

9. $\quad RB = A$ Formula.
$\quad R(96) = 120$ Replace B with 96 and A with 120.

$$\frac{R(96)}{96} = \frac{120}{96}$$ Divide both sides by 96.

$\quad R = 1.25 = 125\%$ Simplify.

10. $\quad c - \dfrac{3}{5} = \dfrac{1}{2}$

$$10\left(c - \frac{3}{5}\right) = 10\left(\frac{1}{2}\right)$$ Multiply both sides by 10 to clear the fractions.

$\quad\quad 10c - 6 = 5$ Multiply using the distributive property.
$\quad 10c - 6 + 6 = 5 + 6$ Add 6 to both sides.
$\quad\quad\quad 10c = 11$ Simplify.

$$\frac{10c}{10} = \frac{11}{10}$$ Divide both sides by 10.

$\quad\quad\quad c = \dfrac{11}{10}$ Simplify.

$\left\{\dfrac{11}{10}\right\}$

11. $\dfrac{x}{-11} = -11$

$-11\left(\dfrac{x}{-11}\right) = -11(-11)$ Multiply both sides by -11.

$\qquad\qquad x = 121$ Simplify.

$\{121\}$

12. $x > 4, \ x \in J$ x is an integer.

13. $RB = A$ Formula.
$0.085B = 51$ Replace R with $8.5\% = 0.085$ and A with 51.

$\dfrac{0.085B}{0.085} = \dfrac{51}{0.085}$ Divide both sides by 0.085.

$B = 600$ Simplify.

14. $6m + 4 < 21$
$6m + 4 - 4 < 21 - 4$ Subtract 4 from both sides.
$6m < 17$ Simplify.

$\dfrac{6m}{6} < \dfrac{17}{6}$ Divide both sides by 6.
     Since 6 is positive, the inequality does not reverse.

$m < \dfrac{17}{6}$ Simplify.

$\left\{m \mid m < \dfrac{17}{6}\right\}$

15. $-\dfrac{2}{3}z = -10$

$-\dfrac{3}{2}\left(-\dfrac{2}{3}z\right) = -\dfrac{3}{2}(-10)$ Multiply both sides by $-\dfrac{3}{2}$.

$z = 15$ Simplify.

$\{15\}$

16. $-11b \leq -143$

$\dfrac{-11b}{-11} \geq \dfrac{-143}{-11}$ Divide by -11. Since -11, is negative, the inequality reverses.

$b \geq 13$

$\{b \mid b \geq 13\}$

Chapter 2 Test

17. $y \le 3$

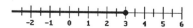

18. $3x - 12 = 100 - 4x$ Add $4x$ to both sides.
    $4x + 3x - 12 = 100 - 4x + 4x$ Simplify.
    $7x - 12 = 100$ Add 12 to both sides.
    $7x - 12 + 12 = 100 + 12$ Simplify.
    $7x = 112$

    $\dfrac{7x}{7} = \dfrac{112}{7}$ Divide both sides by 7.

    $x = 16$ Simplify.

    $\{16\}$

19. $AM + PM = R$
    $AM - AM + PM = R - AM$ Subtract AM from both sides.
    $PM = R - AM$ Simplify.

    $\dfrac{PM}{M} = \dfrac{R - AM}{M}$ Divide both sides by M.

    $P = \dfrac{R - AM}{M}$ Simplify.

20. $18x + 11y - w + 3w - 29y + 13x$
    $= (18 + 13)x + (11 - 29)y + (-1 + 3)w$ Distributive property.
    $= 31x - 18y + 2w$ Add and subtract.

21. $RB = A$ Formula.
    $(0.12)(365) = A$ Replace R with $12\% = 0.12$ and B with 365.
    $43.8 = A$ Multiply.

22. $3x - 5 < 8x + 15$
    $3x - 5 + 5 < 8x + 15 + 5$ Add 5 to both sides.
    $3x < 8x + 20$ Simplify.
    $3x - 8x < 8x + 20$ Subtract $8x$ from both sides.
    $-5x < 20$ Simplify.

    $\dfrac{-5x}{-5} > \dfrac{20}{-5}$ Divide both sides by -5. Since -5 is negative, the inequality reverses.

    $x > -4$ Simplify.

    $\{x \mid x > -4,\ x \in R\}$

23. $8a - 12b + 14c - 12c - 8b - 2a$
    $= (8 - 2)a + (-12 - 8)b + (14 - 12)c$ Distributive property.
    $= 6a - 20b + 2c$ Subtract.

24. $RB = A$
    $R(79.96) = 99.95$ Formula.
    Replace B with 79.96 and A with 99.95.

    $$\frac{R(79.96)}{79.96} = \frac{99.95}{79.96}$$ Divide both sides by 79.96.

    $R = 1.25 = 125\%$ Simplify.

25. $RB = A$ Formula.
    $(0.88)(1.25) = A$ $100\% - 12\% = 88\% = 0.88$.
    $1.10 = A$ Multiply.

    $1.10

Exercises 3.1

CHAPTER 3

EXERCISES 3.1

1. $b^2 \cdot b^3 = b^{2+3}$
   $= b^5$ Add the exponents.

5. $x^6 \cdot x^9 = x^{6+9}$
   $= x^{15}$ Add the exponents.

9. $x^2 \cdot y = x^2 y$ The bases are not the same. The expression $x^2 y$ is in simplest form.

13. $(a^1)^2 = a^{1 \cdot 2}$
    $= a^2$ Multiply the exponents.

17. $(3b)^2 = 3^2 b^2$
    $= 9b^2$ Each factor is raised to the second power.

21. $x^4 \cdot x^5 \cdot x^2 = x^{4+5+2}$
    $= x^{11}$ Add the exponents.

25. $b^{10} \cdot b^8 \cdot b^4 = b^{10+8+4}$
    $= b^{22}$ Add the exponents.

29. $x^4 \cdot x^5 \cdot y^7 \cdot y$
    $= x^{4+5} y^{7+1}$ Add exponents for like bases.
    $= x^9 y^8$

33. $(a^{10})^3 = a^{10 \cdot 3} = a^{30}$ Multiply exponents.

37. $(xy)^0 = 1$ A nonzero base to the zero power equals 1.

41. $(y^2 z^2)^2 = (y^2)^2 (z^2)^2$ Each factor is raised to the second power.
    $= y^{2 \cdot 2} z^{2 \cdot 2}$ Multiply exponents.
    $= y^4 z^4$

45. $y^5 \cdot y^2 \cdot y^3 \cdot y^7 = y^{5+2+3+7}$ Add exponents.
    $= y^{17}$

49. $x^a \cdot x^b = x^{a+b}$ Add exponents.

53. $(x^m)^2 = x^{2 \cdot m} = x^{2m}$ Multiply exponents.

57. $(y^n)^n = y^{n \cdot n} = y^{n^2}$ Multiply exponents.

61. $(15x^2 y z^4)^2 = 15^2 (x^2)^2 y^2 (z^4)^2$ Each factor is raised to the second power.
    $= 225 x^{2 \cdot 2} y^2 z^{4 \cdot 2}$ $15^2 = 225$.
    $= 225 x^4 y^2 z^8$ Multiply exponents.

65. $V = s^3$ Formula.
$V = (3.2)^3$ Replace s with 3.2.
$V = 32.768$ in^3 Volume is measured in cubic units.

69. $V = \pi r^2 h$ Formula.
$V = \pi r^2 (r^3)$ $h = r^3$.
$V = \pi r^5$ Add exponents.
$V = 3.14(3)^5$ Replace π with 3.14 and r with 3.
$V = 763.02$ in^3

73. $A = \dfrac{(B + b)}{2} h$ Formula.

$A = \dfrac{(b^2 + b)}{2} \cdot b^3$ Replace B with b^2 and h with b^3.

$A = \dfrac{(3^2 + 3)}{2} \cdot 3^3$ Replace b with 3.

$A = \dfrac{(9 + 3)}{2} \cdot 27$

$A = 6 \cdot 27$
$A = 162$ in^2 Area is measured in square units.

77. (a.) $A = lw$ Formula.
$A = w^2 w$ $l = w^2$.
$A = w^3$ Add exponents.

(b.) $V = lwh$
$V = w^3 \cdot (w^2)^3$ $lw = w^3$; $h = l^3 = (w^2)^3$.
$V = w^3 \cdot w^6$ Multiply exponents.
$V = w^9$ Add exponents.

81. $2^5 \cdot 2^3 = 2^{5+3}$ Add exponents.
$= 2^8$

85. $-6^0 = -(1)$ $6^0 = 1$.
$= -1$

93. $(-14)^2 - (12)^2 - (-14)(-12)$
$= 196 - 144 - (-14)(-12)$ Evaluate exponents.
$= 196 - 144 - 168$ Multiply.
$= -116$ Add and subtract left to right.

97. $\dfrac{5}{7} + x = \dfrac{5}{21}$

$21\left(\dfrac{5}{7} + x\right) = 21\left(\dfrac{5}{21}\right)$ Multiply both sides by 21 to clear the fractions.

$21\left(\dfrac{5}{7}\right) + 21x = 5$ Distributive property.

Exercises 3.2

$$15 + 21x = 5$$ Simplify.
$$-15 + 15 + 21x = -15 + 5$$ Add -15 to both sides.
$$21x = -10$$ Simplify.
$$\frac{21x}{21} = \frac{-10}{21}$$ Divide both sides by 21.
$$x = -\frac{10}{21}$$ Simplify.

$$\left\{-\frac{10}{21}\right\}$$

EXERCISES 3.2

1. $6^{-1} = \dfrac{1}{6^1}$ The meaning of a negative exponent.

   $= \dfrac{1}{6}$ $6^1 = 6$.

5. $z^{-4} = \dfrac{1}{z^4}$ The meaning of a negative exponent.

9. $\dfrac{x^7}{x^{10}} = x^{7-10}$ Subtract the exponents.

   $= x^{-3}$

   $= \dfrac{1}{x^3}$ The meaning of a negative exponent.

13. $(x^{-2})^2 = x^{-2 \cdot 2}$ Multiply the exponents.
    $= x^{-4}$

    $= \dfrac{1}{x^4}$ The meaning of a negative exponent.

17. $\left(\dfrac{a^{-3}}{b}\right)^2 = \dfrac{(a^{-3})^2}{b^2}$ Raise both the numerator and denominator to the 6second power.

    $= \dfrac{a^{-6}}{b^2}$ Multiply the exponents.

    $= \dfrac{1}{a^6 b^2}$ The meaning of a negative exponent.

21. $10^{-4} \cdot 10^2 = 10^{-4+2}$ Add the exponents.
    $= 10^{-2}$

$$= \frac{1}{10^2}$$ The meaning of a negative exponent.

$$= \frac{1}{100}$$ Simplify.

25. $y^{-5} \cdot y^{13} \cdot y^{-8} = y^{-5+13+(-8)}$ Add the exponents.
$$= y^0$$ A nonzero base to the zero power equals 1.
$$= 1$$

29. $(y^4 \cdot y^{-5})^{-1} = (y^{-1})^{-1}$ Add the exponents inside the parentheses.
$$= y^1$$ Multiply the exponents.
$$= y$$

33. $\dfrac{x^a}{x^n} = x^{a-n}$ Subtract the exponents.

37. $\left(\dfrac{x^5 y^2}{w^2 z^3}\right)^3 = \dfrac{(x^5 y^2)^3}{(w^2 z^3)^3}$ Raise a quotient to a power.

$$= \frac{(x^5)^3 (y^2)^3}{(w^2)^3 (z^3)^3}$$ Raise a product to a power.

$$= \frac{x^{15} y^6}{w^6 z^9}$$ Multiply the exponents.

41. $\left(\dfrac{4^3 a^{-3}}{4 a^{-4}}\right)^{-2} = \dfrac{(4^3 a^{-3})^{-2}}{(4 a^{-4})^{-2}}$ Raise a quotient to a power.

$$= \frac{(4^3)^{-2}(a^{-3})^{-2}}{(4)^{-2}(a^{-4})^{-2}}$$ Raise a product to a power.

$$= \frac{4^{-6} a^6}{4^{-2} a^8}$$ Multiply the exponents.

$$= 4^{-6-(-2)} a^{6-8}$$ Subtract the exponents.
$$= 4^{-4} a^{-2}$$

$$= \frac{1}{4^4} \cdot \frac{1}{a^2}$$ The meaning of a negative exponent.

$$= \frac{1}{256 a^2}$$ Simplify.

45. $\dfrac{3^0 \cdot 3^{-7}}{3^{-5}} = \dfrac{3^{-7}}{3^{-5}}$ Add the exponents in the numerator.

$$= 3^{-7-(-5)}$$ Subtract the exponents.
$$= 3^{-2}$$

Exercises 3.2

$$= \frac{1}{3^2}$$ The meaning of a negative exponent.

$$= \frac{1}{9}$$ Simplify.

49. $\dfrac{x^3 \cdot x^{-2}}{x^{-5} \cdot x^4} = \dfrac{x^1}{x^{-1}}$ Add the exponents.

$$= x^{1-(-1)}$$ Subtract the exponents.
$$= x^2$$ Simplify.

53. $\left(\dfrac{a^7 \cdot a^{-3}}{a^{-6} \cdot a^4}\right)^{-2} = \dfrac{(a^7 \cdot a^{-3})^{-2}}{(a^{-6} \cdot a^4)^{-2}}$ Raise a quotient to a power.

$$= \frac{(a^7)^{-2}(a^{-3})^{-2}}{(a^{-6})^{-2}(a^4)^{-2}}$$ Raise a product to a power.

$$= \frac{a^{-14} \cdot a^6}{a^{12} \cdot a^{-8}}$$ Multiply the exponents.

$$= \frac{a^{-8}}{a^4}$$ Add the exponents.

$$= a^{-8-4}$$ Subtract the exponents.
$$= a^{-12}$$

$$= \frac{1}{a^{12}}$$ The meaning of a negative exponent.

57. $\dfrac{x^2 \cdot x^{-3} \cdot x^5}{x^4 \cdot x^0} = \dfrac{x^{2+(-3)+5}}{x^{4+0}}$ Add the exponents.

$$= \frac{x^4}{x^4}$$

$$= x^{4-4}$$ Subtract the exponents.
$$= x^0$$
$$= 1$$ A nonzero base raised to the zero power equals 1.

61. $\dfrac{x^{31} x^{-22}}{x^{-16} x^{24}} = \dfrac{x^{31+(-22)}}{x^{-16+24}}$ Add the exponents.

$$= \frac{x^9}{x^8}$$

$$= x^{9-8}$$ Subtract the exponents.
$$= x^1$$
$$= x$$

65. $\left(\dfrac{x^7y^8}{x^4y^9}\right)^{-3} = \dfrac{(x^7y^8)^{-3}}{(x^4y^9)^{-3}}$ Raise a quotient to a power.

$= \dfrac{(x^7)^{-3}(y^8)^{-3}}{(x^4)^{-3}(y^9)^{-3}}$ Raise a product to a power.

$= \dfrac{x^{-21}y^{-24}}{x^{-12}y^{-27}}$ Multiply the exponents.

$= x^{-21-(-12)}y^{-24-(-27)}$ Subtract the exponents.
$= x^{-9}y^3$

$= \dfrac{y^3}{x^9}$ The meaning of a negative exponent.

69. $5 \times 10^{-9} = 0.000000005$ in.   Move the decimal 9 places to the left.

73. $V = IR$ Formula.
$V = (10^{-3})(10^4)$ Replace I with 10^{-3} and R with 10^4.

$V = 10^1$ Add the exponents.
$V = 10$

77. $A = wl$ Formula.

$A = \dfrac{1}{l} \cdot l$ $w = \dfrac{1}{l}$.

$A = l^{1-1}$ Subtract the exponents.
$A = l^0$
$A = 1$ A nonzero base to the zero power equals 1.

81. One possible answer: x^3y^{-2} indicates multiplication of x^3 and y^{-2}; $x^3 + y^{-2}$ indicates addition of x^3 and y^{-2}.

85. $\dfrac{4^{-2} \cdot 2^{3a+1} \cdot 8^a}{16^{2a}}$

$= \dfrac{(2^2)^{-2} \cdot 2^{3a+1} \cdot (2^3)^a}{(2^4)^{2a}}$ $4 = 2^2;\ \ 8 = 2^3;\ \ 15 = 2^4$.

$= \dfrac{2^{-4} \cdot 2^{3a+1} \cdot 2^{3a}}{2^{8a}}$ Multiply the exponents.

$= \dfrac{2^{-4+3a+1+3a}}{2^{8a}}$ Add the exponents in the numerator.

$= \dfrac{2^{6a-3}}{2^{8a}}$

$= 2^{6a-3-8a}$ Subtract the exponents.

Exercises 3.3

$\quad\quad = 2^{-2a-3}$
$\quad\quad = 2^{-(2a+3)}$ Distributive property.

$\quad\quad = \dfrac{1}{2^{2a+3}}$ The meaning of a negative exponent.

89. $\;-5[3(-4) - (-4)(-9)] + (-5)(15)$
$\;= -5(-12 - 36) + (-5)(15)$ Multiply inside the brackets.
$\;= -5(-48) + (-5)(15)$ Subtract inside the parentheses.
$\;= 240 - 75$ Multiply.
$\;= 165$ Subtract.

93. $\;-82y = 3936$

$\quad\dfrac{-82y}{-82} = \dfrac{3936}{-82}$ Divide both sides by -82.

$\quad\quad y = -48$ Simplify.

$\{-48\}$

EXERCISES 3.3

1. 50,000 Original number.

   5.0000 Step 1: Write a number between 1 and 10. Move the decimal four places left.

   10^4 Step 2: The exponent of 10 is 4 since $5.0 \times 10^4 = 50000$.

   $50,000 = 5.0 \times 10^4$ Step 3: Write the original number as the product of the number in Step 1 and the power of 10 in Step 2.

5. 9,000,000 Original number.

   9.000000 Step 1: Move the decimal 6 places left.

   10^6 Step 2: The exponent of 10 is 6 since $9.0 \times 10^6 = 9000000$.

   $9,000,000 = 9.0 \times 10^6$ Step 3: Write the original number as the product of the number in Step 1 and the power of 10 in Step 2.

9. $9.3 \times 10^2 = 930$ To multiply by 10^2 move the decimal point two places to the right.

13. $5.4 \times 10^7 = 54,000,000$ To multiply by 10^7, move the decimal point seven places to the right.

17. 377,000 Original number.

    3.77000 Step 1: Move the decimal five places left.

10^5

$377,000 = 3.77 \times 10^5$

Step 2: The exponent of 10 is 5 since $3.77 \times 10^5 = 377,000$.

Step 3: Write the original number as the product of the number in Step 1 and the power of 10 in Step 2.

21. 611,000,000

6.11000000

10^8

$611,000,000 = 6.11 \times 10^8$

Original number.

Step 1: Move the decimal eight places left.

Step 2: The exponent of 10 is 8 since $6.11 \times 10^8 = 611,000,000$.

Step 3: Write the original number as the product of the number in Step 1 and the power of 10 in Step 2.

25. 0.0000922

00009.22

10^{-5}

$0.0000922 = 9.22 \times 10^{-5}$

Original number.

Step 1: Move the decimal five places right.

Step 2: The exponent of 10 is -5 since $9.22 \times 10^{-5} = 0.0000922$.

Step 3: Write the original number as the product of the number in Step 1 and the power of 10 in Step 2.

29. $6.89 \times 10^4 = 68900$

To multiply by 10^4, move the decimal point four places to the right.

33. $4.6 \times 10^5 = 460,000$

To multiply by 10^5, move the decimal point five places to the right.

37. 3784

3.784

10^3

$3784 = 3.784 \times 10^3$

Original number.

Step 1: Move the decimal three places left.

Step 2: The exponent of 10 is 3 since 3.784×10^3 is 3784.

Step 3: Write the original number as the product of the number in Step 1 and the power of 10 in Step 2.

41. 0.000484

0004.84

10^{-4}

$0.000484 = 4.84 \times 10^{-4}$

Original number.

Step 1: Move the decimal four places right.

Step 2: The exponent of 10 is -4 since $4.84 \times 10^{-4} = 0.000484$.

Step 3: Write the original number as the product of the number in Step 1 and the power of 10 in Step 2.

45. 296,200,000

2.96200000

Original number.

Step 1: Move the decimal eight places left.

Exercises 3.3

	10^8	Step 2: The exponent of 10 is 8 since $2.962 \times 10^8 = 296{,}200{,}000$.
	$296{,}200{,}000 = 2.962 \times 10^8$	Step 3: Write the original number as the product of the number in Step 1 and the power of 10 in Step 2.

49. $2.36 \times 10^{-9} = 0.00000000236$ To multiply by 10^{-9}, move the decimal point nine places to the left.

53. $9.11 \times 10^{-8} = 0.0000000911$ To multiply by 10^{-8}, move the decimal point eight places to the left.

57. 5,870,000,000,000   Original number.

    5.870000000000   Step 1:

    10^{12} Step 2. The decimal was moved twelve places left.

    $5{,}870{,}000{,}000{,}000 = 5.87 \times 10^{12}$ mi   Step 3.

61. 0.003   Original number.

    0003.   Step 1.

    10^3 Step 2: The decimal was moved three places right.

    $0.003 = 3.0 \times 10^{-3}$ Step 3.

65. $2.55 \times 10^{13} = 25{,}500{,}000{,}000{,}000$ mi   To multiply by 10^{13}, move the decimal point thirteen places to the right.

69. $1.2 \times 10^7 = 12{,}000{,}000$ people   To multiply by 10^7, move the decimal point seven places to the right.

73. $\dfrac{(3.25 \times 10^{-2})(2.4 \times 10^2)}{(4.8 \times 10^{-2})(2.5 \times 10^{-2})}$

   $= \dfrac{(3.25)(2.4)(10^{-2} \cdot 10^2)}{(4.8)(2.5)(10^{-2} \cdot 10^{-2})}$ Rearrange the factors to get the powers of 10 together.

   $= \dfrac{(7.8)(10^{-2+2})}{(12)(10^{-2+(-2)})}$ Multiply. Add the exponents.

   $= \dfrac{(7.8)(10^0)}{(12)(10^{-4})}$ Simplify.

   $= (0.65)(10^{0-(-4)})$ Divide. Subtract the exponents.
   $= 0.65 \times 10^4$ Simplify.
   $= 6.5 \times 10^3$ Move the decimal point one place to the right.

81. $RB = A$
$R(690) = 483$ Formula.
690 − 207 = 483 seats not in the balcony.

$$\frac{R(690)}{690} = \frac{483}{690}$$

Divide both sides by 690.

$R = 0.7 = 70\%$ Simplify.

85. $\dfrac{x}{-3} - 11 = 14$

$-3\left(\dfrac{x}{-3} - 11\right) = -3(14)$ Multiply both sides by −3 to clear the fractions.

$-3\left(\dfrac{x}{-3}\right) - 3(-11) = -42$ Distributive property.

$x + 33 = -42$ Simplify.
$x + 33 - 33 = -42 - 33$ Subtract 33 from both sides.
$x = -75$

$\{-75\}$

EXERCISES 3.4

1. coefficient: 5
   degree: 2

5. coefficient: 16
   degree: 4

9. coefficient: $\dfrac{14}{3}$
   degree: 2

13. Monomial   One term.

17. coefficient: −14
    degree: 10   Add the exponents of the variables.

21. coefficient: −3
    degree: 9   Add the exponents of the variables.

25. coefficient: $-\dfrac{15}{2}$
    degree: 0   The degree of a constant is 0.

29. Monomial   One term.

33. degree: 3   Same as degree of term with highest degree.

37. degree: 5   Same as degree of term with highest degree. Add the exponents of the variables.

Exercises 3.4

41. degree: 14

Same as degree of $-14a^7b^7$. Add the exponents of the variables.

45. $-14x^5 - 3x^2 + 7x - 6$

Terms are written in decreasing degree reading from left to right.

49. (a.) Jay

Degree 4.

(b.) 3

(c.) $7x^3 - 4x^2 + 2x$

Terms are written in decreasing degree reading from left to right.

(d.) Melissa

Two terms.

(e.) Mollie

Three terms.

(f.) Jay: $5x^4 = 5(3)^4 = 5(81) = 405$ hr

Replace x with 3.

Mollie: $7x^3 - 4x^2 + 2x$
$= 7(3)^3 - 4(3)^2 + 2(3)$
$= 7(27) - 4(9) + 6$
$= 189 - 36 + 6$
$= 159$ hr

Replace x with 3.

Matt: $2y^2 = 2(2)^2 = 2(4) = 8$ hr

Replace y with 2.

Melissa: $4y^2 + 5y = 4(2)^2 + 5(2)$
$= 4(4) + 10$
$= 16 + 10$
$= 26$ hr

Replace y with 2.

Joe: $3x^2y = 3(3)^2(2)$
$= 3(9)(2)$
$= 54$ hr

Replace x with 3 and y with 2.

53. One possible answer: The degree of a term is the sum of the exponents of the variables in that term.

57. 2 terms
    Binomial

$7x^2$; $(2x + 3)(3x - 4)$

61. $RB = A$ Formula.
    $R(134) = 13$ Replace B with 134 and A with 13.

$$\frac{R(134)}{134} = \frac{13}{134}$$ Divide both sides by 134.

$R = 0.097 = 9.7\%$ Simplify.

65. $ax + b = c$
    $24x + 56 = -20$ Replace a with 24, b with 56 and c with -20.
    $24x + 56 - 56 = -20 - 56$ Subtract 56 from both sides.
    $24x = -76$ Simplify.

$$\frac{24x}{24} = \frac{-76}{24}$$ Divide both sides by 24.

$$x = -\frac{19}{6}$$ Simplify.

$\left\{-\frac{19}{6}\right\}$

69. $-1.2y + 6.54 = -7.86$
    $-1.2y + 6.54 - 6.54 = -7.86 - 6.54$ Subtract 6.54 from both sides.
    $-1.2y = -14.4$ Simplify.

$$\frac{-1.2y}{-1.2} = \frac{-14.4}{-1.2}$$ Divide both sides by -1.2.

$y = 1.2$ Simplify.

$\{12\}$

EXERCISES 3.5

1. $(5a - 2b) + (3a - b)$
   $= (5a + 3a) + (-2b - b)$ Group like terms.
   $= (5a + 3a) + [-2b + (-b)]$ Change subtraction to addition.
   $= 8a - 3b$ Combine like terms.

5. $(7m - 8) + (3m - 9)$
   $= (7m + 3m) + (-8 - 9)$ Group like terms.
   $= (7m + 3m) + [-8 + (-9)]$ Change subtraction to addition.
   $= 10m - 17$ Combine like terms.

9. $(12a - 9b) - (16a + 3b)$
   $= (12a - 9b) + (-16a - 3b)$ Change to addition.
   $= (12a - 16a) + (-9b + (-3b)$ Group like terms.
   $= [12a + (16a)] + [-9b + (-3b)]$ Change subtraction to addition.
   $= -4a - 12b$ Combine like terms.

13. $3a - 6a - 9a = -24$
               $-12a = -24$ Combine terms.
                     $a = 2$ Divide both sides by -12.
    $\{2\}$

17. $(x^2 - 8x + 3) + (4x^2 + 9x + 7) + (-8x^2 + 6x - 11)$
    $= (x^2 + 4x^2 - 8x^2) + (-8x + 9x + 6x) + (3 + 7 - 11)$ Group like terms.
    $= [x^2 + 4x^2 + (-8x^2)] + (-8x + 9x + 6x) + [3 + 7 + (-11)]$ Change to addition.
    $= -3x^2 + 7x - 1$ Combine like terms.

21. $15y^2 - 8y + 19$
      $\underline{12y^2 - 11y - 31}$
      $27y^2 - 19y - 12$ Combine like terms.

25. $a^2 - 7ab + 3b^2$
      $\underline{7a^2 + 3ab - 9b^2}$
      $8a^2 - 4ab - 6b^2$ Combine like terms.

29. $(21y^2 + 19y - 28) - (-8y^2 - 7y + 15)$
    $= 21y^2 + 19y - 28) + (8y^2 + 7y - 15)$ Change to addition.
    $= (21y^2 + 8y^2) + (19y + 7y) + (-28 - 15)$ Group like terms.
    $= (21y^2 + 8y^2) + (19y + 7y) + [-28 + (-15)]$ Change to addition.
    $= 29y^2 + 26y - 43$ Combine like terms.

33. $8x - 3y + 8$
      $\underline{7x + 6y + 3}$

      $8x - 3y + 8$
      $\underline{-7x + 6y + 3}$ Subtract by adding the opposite.
        $x - 9y + 5$

37. $(8x + 3) + (-5x + 2) = -2x - 15$
                      $3x + 5 = -2x - 15$ Combine terms.
                            $3x = -2x - 20$ Subtract 5 from both sides.

$$5x = -20$$
$$x = -4$$

Add $2x$ to both sides.
Divide both sides by 5.

$\{-4\}$

41. $(-6x - 10) - (2x - 6) = 4 - 2x$
$-6x - 10 - 2x + 6 = 4 - 2x$
$-8x - 4 = 4 - 2x$
$-6x - 4 = 4$
$-6x = 8$

Remove the parentheses by adding the opposite.
Combine terms.
Add $2x$ to both sides.
Add 4 to both sides.

$$x = -\frac{4}{3}$$

Divide both sides by -6.

$\left\{-\dfrac{4}{3}\right\}$

45. $(0.25x - 2y + 1.5z) + (1.25x - 0.75y + z)$
$\quad + (-0.5x + 0.5y - 0.5z)$
$= (0.25x + 1.25x - 0.5x) + (-2y - 0.75y + 0.5y)$
$\quad + (1.5z + z - 0.5z)$
$= [0.25x + 1.25x + (-0.5x)] + [-2y + (-0.75y) + 0.5y]$
$\quad + [1.5z + z + (-0.5z)]$
$= x - 2.25y + 2z$

Group like terms.

Change to addition.

Combine like terms.

49. $(5x - 2) - [-5x - (x - 12) - 4] + 6$
$= (5x - 2) - (-5x - x + 12 - 4) + 6$
$= (5x - 2) - (-6x + 8) + 6$
$= 5x - 2 + 6x - 8 + 6$
$= (5x + 6x) + [-2 + (-8) + 6]$
$= 11x - 4$

Change to addition inside brackets.
Combine like terms inside the parentheses.
Change to addition.
Group like terms.
Combine like terms.

53. $(3x + 8) - (5 - 2x) = (4x + 2) + (-3x + 5)$
$(3x + 8) + (-5 + 2x) = (4x + 2) + (-3x + 5)$
$5x + 3 = x + 7$
$4x + 3 = 7$
$4x = 4$
$x = 1$

Change to addition.
Combine terms.
Subtract x from both sides.
Subtract 3 from both sides.
Divide both sides by 4.

$\{1\}$

57. number: x

$(x + 4) + (x - 17) = 75$
$2x - 13 = 75$
$2x = 88$
$x = 44$

$x + 4$: four more than a number.
$x - 17$: seventeen less than the number.
Combine like terms.
Add 13 to both sides.
Divide both sides by 2.

The number is 44.

61. number: x

$(4x + 6) + (8x - 12) = 2x - 31$

$12x - 6 = 2x - 31$
$12x = 2x - 25$

$4x + 6$: six more than four times a number.
$8x - 12$: twelve less than eight times the number.
$2x - 31$: thirty-one less than twice the number.
Combine like terms.
Add 6 to both sides.

Exercises 3.5

$10x = -25$ Subtract $2x$ from both sides.

$x = -\dfrac{5}{2}$ Divide both sides by 10.

The number is $-\dfrac{5}{2}$.

65. $x + (x + 6) + (x + 2) = 41$ Add the three side lengths.
$3x + 8 = 41$ Combine like terms.
$3x = 33$ Subtract 8 from both sides.
$x = 11$ Divide both sides by 3.

$x = 11$
$x + 6 = 11 + 6 = 17$
$x + 2 = 11 + 2 = 13$

The side lengths are 11 ft, 17 ft and 13 ft.

69. smaller number: x
larger number: $5 + 2x$ Five more than twice the smaller.

$(5 + 2x) - 3x = -8$ "Difference" indicates subtraction.
$-x + 5 = -8$ Combine like terms.
$-x = -13$ Subtract 5 from both sides.
$x = 13$ Divide both sides by -1.

$x = 13$
$5 + 2x = 5 + 2(13) = 31$

The numbers are 13 and 31.

73. Minh's score: x
Cindy's score: $4 + 5x$ Four more than five times Minh's score.

$(4 + 5x) - x = 4900$ "Difference" indicates subtraction.
$4 + 4x = 4900$ Combine like terms.
$4x = 4896$ Subtract 4 from both sides.
$x = 1224$ Divide both sides by 4.

$x = 1224$
$4 + 5x = 4 + 5(1224) = 6124$

Minh's score: 1224; Cindy's score: 6124

77. $I_2 - I_1 = 60$
$(28 + 15t - t^2) - (16 - 3t - t^2) = 60$ Replace I_2 with $28 + 15t - t^2$ and I_1 with $16 - 3t - t^2$.

$(28 + 15t - t^2) + (-16 + 3t + t^2) = 60$ Change to addition.
$12 + 18t = 60$ Combine like terms.
$18t = 48$ Subtract 12 from both sides.

$t = \dfrac{8}{3}$ Divide both sides by 18.

$\dfrac{8}{3} = 2\dfrac{2}{3}$ sec

81. One possible answer: When subtracting polynomials there is an extra step. First you change the subtraction to addition by adding the opposite.

85. $-4y^2 + 3y - 6 + ? = 0$ Let ? represent the unknown algebraic expression.
      $3y - 6 + ? = 4y^2$ Add $4y^2$ to both sides.
      $-6 + ? = 4y^2 - 3y$ Subtract $3y$ from both sides.
      $? = 4y^2 - 3y + 6$ Add 6 to both sides.

The expressions are opposites.

89. $(ab^2)^{-1} = \dfrac{1}{(ab^2)^1}$ The meaning of a negative exponent.

    $= \dfrac{1}{ab^2}$

93. $\dfrac{x^{-3}}{x^{-5}} = x^{-3-(-5)}$ Subtract the exponents.

    $= x^2$ Simplify.

97. Look at the pattern in the "save" column. Each is a power of 2 and the exponent matches the day. Hence, on 8/17 he will save 2^{17} cents or 131072 cents = $1310.72.

EXERCISES 3.6

1.

	Rate ·	Time =	Distance
1st car	65	t	65t
2nd car	58	t	58t

$\begin{pmatrix}\text{distance traveled}\\ \text{by 1st car}\end{pmatrix} - \begin{pmatrix}\text{distance traveled}\\ \text{by 2nd car}\end{pmatrix} = \begin{pmatrix}\text{distance}\\ \text{between the cars}\end{pmatrix}$

$65t - 58t = 420$
$7t = 420$
$t = 60$

60 hr

5.

	Rate ·	Time =	Distance
1st cyclist	6	t	6t
2nd cyclist	9	t	9t

$\begin{pmatrix}\text{distance traveled}\\ \text{by 1st cyclist}\end{pmatrix} + \begin{pmatrix}\text{distance traveled}\\ \text{by 2nd cyclist}\end{pmatrix} = \begin{pmatrix}\text{distance between}\\ \text{the two cyclists}\end{pmatrix}$

Exercises 3.6

$6t - 9t = 24$ — Substitute information from chart.
$15t = 24$ — Combine like terms.

$t = \dfrac{8}{5}$ — Divide both sides by 15.

$\dfrac{8}{5} = 1.6$ hr

9.

	Value of Each Coin	·	Number of Coins	=	Value of Coins
Quarters	0.25		n		$0.25n$
Half dollars	0.50		n		$0.50n$

$\begin{pmatrix} \text{Value of} \\ \text{Quarters} \end{pmatrix} + \begin{pmatrix} \text{Value of} \\ \text{Half dollars} \end{pmatrix} = \begin{pmatrix} \text{Total} \\ \text{Value} \end{pmatrix}$

$0.25n + 0.50n = 101.25$ — Substitute information from chart.
$0.75n = 101.25$ — Combine like terms.
$n = 135$ — Divide both sides by 0.75.

135 of each coin.

13.

	Rate	Principle	Interest
1st Nat'l	$6.75\% = 0.0675$	m	$0.0675m$
2nd Nat'l	$7.15\% = 0.0715$	m	$0.0715m$

$\begin{pmatrix} \text{Interest from} \\ \text{2}^{nd} \text{ Nat'l} \end{pmatrix} - \begin{pmatrix} \text{Interest from} \\ \text{1}^{st} \text{ Nat'l} \end{pmatrix} = \begin{pmatrix} \text{Difference} \\ \text{between interests} \end{pmatrix}$

$0.0715m - 0.0675m = 10.24$ — Substitute from chart.
$0.004m = 10.24$ — Combine like terms.
$m = 2560$ — Divide both sides by 0.004.

$2560 in each bank.

17. Sally's salary: $2x$ — Twice Fred's salary.
    Fred's salary: x
    Joy's salary: $2x - 50$ — 50 less than Sally's salary.

$2x + x + 2x - 50 = 3950$ — Sum of all three salaries = 3950.
$5x - 50 = 3950$ — Combine like terms.
$5x = 4000$ — Add 50 to both sides.
$x = 800$ — Divide both sides by 5.

$x = 800$
$2x = 2(800) = 1600$
$2x - 50 = 2(800) - 50 = 1550$

Sally: $1600; Fred: $800; Joy: $1550

21. One possible answer: First you must read the problem several times slowly and carefully.

25. Jill's portion: x
    Mike's portion: $0.75x$ 25% less indicates Mike's share is 75% of Jill's share.

$$x + 0.75x = 56000$$
$$1.75x = 56000$$
$$x = 32000$$
$$0.75x = 0.75(32000) = 24000$$

Sum of shares equals total profit.
Combine like terms.
Divide both sides by 1.75.

Jill: $32,000; Mike: $24,000

29. True   −82 is to the left of 114 on the number line.

33.
$$5x - 17 > 32$$
$$5x - 17 + 17 > 32 + 17$$
$$5x > 49$$

Add 17 to both sides.

$$\frac{5x}{5} > \frac{49}{5}$$

Divide both sides by 5.

$$x > 9.8$$

$\{x \mid x > 9.8\}$

37. number of sacks: x

$$210 + 75x \leq 1500$$
$$-210 + 210 + 75x \leq -210 + 1500$$
$$75x \leq 1290$$

Joe's weight plus wheat's weight is less than or equal to 1500.
Add −210 to both sides.

$$\frac{75x}{75} \leq \frac{1290}{75}$$

Divide both sides by 75.

$$x \leq 17.2$$

At most 17 sacks.

EXERCISES 3.7

1. $5(2x - 3) = 5(2x) + 5(-3)$
   $= 10x + (-15)$
   $= 10x - 15$

   Use the distributive property.
   Simplify.
   Change to subtraction.

5. $7c(4a - 3b + 4c) = 7c(4a) + 7c(-3b) + 7c(4c)$
   $= 28ac + (-21bc) + 28c^2$
   $= 28ac - 21bc + 28c^2$

   Use the distributive property.
   Simplify.
   Change to subtraction.

9. $(x - 1)(x - 2) = (x - 1)(x) + (x - 1)(-2)$
   $= x^2 - x - 2x + 2$
   $= x^2 - 3x + 2$

   Distributive property.
   Multiply.
   Combine like terms.

Exercises 3.7

13. $(c - 4)(c + 1) = (c - 4)(c) + (c - 4)(1)$ Distributive property.
$ = c^2 - 4c + c - 4$ Multiply.
$ = c^2 - 3c - 4$ Combine like terms.

17. $(y + 5)(y + 2) = (y + 5)(y) + (y + 5)(2)$ Distributive property.
$ = y^2 + 5y + 2y + 10$ Multiply.
$ = y^2 + 7y + 10$ Combine like terms.

21. $-5(2a - 4) - 3(4a + 4) = -14$
$-10a + 20 - 12a - 12 = -14$ Distributive property.
$-22a + 8 = -14$ Combine terms.
$-22a = -22$ Subtract 8 from both sides.
$a = 1$ Divide both sides by -22.
{1}

25. $(y - 2)(y^2 + 4y - 2)$
$= (y - 2)(y^2) + (y - 2)(4y) + (y - 2)(-2)$ Distributive property.
$= y^3 - 2y^2 + 4y^2 - 8y - 2y + 4$ Multiply.
$= y^3 + 2y^2 - 10y + 4$ Combine like terms.

29. $(a + b)(a - b)(a - b)$
$= [(a + b)(a) + (a + b)(-b)](a - b)$ Use the distributive property on $(a + b)(a - b)$.
$= (a^2 + ab - ab - b^2)(a - b)$ Multiply inside brackets.
$= (a^2 - b^2)(a - b)$ Combine like terms inside parentheses.
$= (a^2 - b^2)(a) + (a^2 - b^2)(-b)$ Distributive property.
$= a^3 - ab^2 - a^2b + b^3$ Multiply.
$= a^3 - a^2b - ab^2 + b^3$ Rearrange terms so the exponents of a are in descending order.

33. $(2x + 3)(x^2 - 5x - 6)$
$= (2x + 3)(x^2) + (2x + 3)(-5x) + (2x + 3)(-6)$ Distributive property.
$= 2x^3 + 3x^2 - 10x^2 - 15x - 12x - 18$ Multiply.
$= 2x^3 - 7x^2 - 27x - 18$ Combine like terms.

37.
$$\begin{array}{r} 4x^3 + 3xy + 2 \\ \underline{x - y} \\ -4x^3y - 3xy^2 - 2y \\ \underline{4x^4 + 3x^2y + 2x } \\ 4x^4 + 3x^2y + 2x - 4x^3y - 3xy^2 - 2y \end{array}$$
Multiply by $-y$.
Multiply by x.
Add.

41. $(a - 3b)(4a + b)(2a - b)$
$= [(a - 3b)(4a) + (a - 3b)(b)](2a - b)$ Use the distributive property on $(a - 3b)(4a + b)$.
$= (4a^2 - 12ab + ab - 3b^2)(2a - b)$ Multiply inside brackets.
$= (4a^2 - 11ab - 3b^2)(2a - b)$ Combine like terms inside parentheses.
$= (4a^2 - 11ab - 3b^2)(2a)$
$ + (4a^2 - 11ab - 3b^2)(-b)$ Distributive property.
$= 8a^3 - 22a^2b - 6ab^2 - 4a^2b + 11ab^2 + 3b^3$ Multiply.
$= 8a^3 - 26a^2b + 5ab^2 + 3b^3$ Combine like terms.

45. $3x(4x - 5) - 12 = 6x(2x - 3)$
$12x^2 - 15x - 12 = 12x^2 - 18x$ Distributive property.
$-15x - 12 = -18x$ Subtract $12x^2$ from both sides.

$$-12 = -3x$$
$$4 = x$$

{4}

Add $15x$ to both sides.
Divide both sides by -3.

49. $(2a - b)(4a^2 + 2ab + b^2)$
$= (2a - b)(4a^2) + (2a - b)(2ab) + (2a - b)(b^2)$
$= 8a^3 - 4a^2b + 4a^2b - 2ab^2 + 2ab^2 - b^3$
$= 8a^3 - b^3$

Distributive property.
Multiply.
Combine like terms.

53. $(0.6x^2 + 0.5x + 0.3)(0.2x - 0.3)$
$= (0.6x^2 + 0.5x + 0.3)(0.2x) + (0.6x^2 + 0.5x + 0.3)(-0.3)$
$= 0.12x^3 + 0.1x^2 + 0.06x - 0.18x^2 - 0.15x - 0.09$
$= 0.12x^3 - 0.08x^2 - 0.09x - 0.09$

Distributive property.
Multiply.
Combine like terms.

57. $(x + y + z)(x + y - z)$
$= (x + y + z)(x) + (x + y + z)(y) + (x + y + z)(-z)$
$= x^2 + xy + xz + xy + y^2 + yz - xz - yz - z^2$
$= x^2 + 2xy + y^2 - z^2$

Distributive property.
Multiply.
Combine like terms.

61. $(-x^3 - 5)(2x^2 + 5x + 4)$
$= (-x^3 - 5)(2x^2) + (-x^3 - 5)(5x) + (-x^3 - 5)(4)$
$= -2x^5 - 10x^2 - 5x^4 - 25x - 4x^3 - 20$
$= -2x^5 - 5x^4 - 4x^3 - 10x^2 - 25x - 20$

Distributive property.
Multiply.
Write with exponents in descending order.

65.
$$(x + 4)(x^2 - 4x + 16) = x^3 - 5x + 9$$
$$(x + 4)(x^2) + (x + 4)(-4x) + (x + 4)(16) = x^3 - 5x + 9$$
$$x^3 + 4x^2 - 4x^2 - 16x + 16x + 64 = x^3 - 5x + 9$$
$$x^3 + 64 = x^3 - 5x + 9$$
$$64 = -5x + 9$$
$$55 = -5x$$
$$-11 = x$$

Distributive property.
Multiply.
Combine like terms.
Subtract x^3 from both sides.
Subtract 9 from both sides.
Divide both sides by -5.

{-11}

69. number: x

$$2x - (x - 4)(3) = 7$$
$$2x - 3x + 12 = 7$$
$$-x + 12 = 7$$
$$-x = -5$$
$$x = 5$$

The number is 5.

$x - 4$: Four less than a number.
$2x$: Twice the number.
Distributive property.
Combine like terms.
Subtract 12 from both sides.
Divide both sides by -1.

73. Sean's age: x
Gwen's age: $x + 5$

$$x + x + 5 = 84$$
$$2x + 5 = 84$$
$$2x = 79$$
$$x = 39\frac{1}{2}$$

5 years older.

Sum indicates addition.
Combine like terms.
Subtract 5 from both sides.
Divide by 2.

Exercises 3.7

$x + 5 = 39\frac{1}{2} + 5 = 44\frac{1}{2}$

Gwen is $44\frac{1}{2}$ years old.

77. One possible answer: The degree of the product will be the same as the degree of the term resulting from the product of the highest degree terms of the two polynomials. When multiplying these terms you add the exponents of like bases.

81. $(x - 2)(3x^n - 4x^{n-1})$
$= (x - 2)(3x^n) + (x - 2)(-4x^{n-1})$ Distributive property.
$= 3x^{n+1} - 6x^n - 4x^{n-1+1} + 8x^{n-1}$ Multiply. Add the exponents of like bases.
$= 3x^{n+1} - 6x^n - 4x^n + 8x^{n-1}$
$= 3x^{n+1} - 10x^n + 8x^{n-1}$ Combine like terms.

85. $\left(\dfrac{x^2 \cdot y^7}{x^{-4} \cdot y^6}\right)^{-2} = \dfrac{(x^3 \cdot y^7)^{-2}}{(x^{-4} \cdot y^6)^{-2}}$ Raise a quotient to a power.

$= \dfrac{(x^3)^{-2}(y^7)^{-2}}{(x^{-4})^{-2}(y^6)^{-2}}$ Raise a product to a power.

$= \dfrac{x^{-6} y^{-14}}{x^8 y^{-12}}$ Multiply the exponents.

$= x^{-6-8} y^{-14-(-12)}$ Subtract the exponents.
$= x^{-14} y^{-2}$ Simplify.

$= \dfrac{1}{x^{14} y^2}$ The meaning of a negative exponent.

89. 123,000   Original number.

1.23000   Step 1: Move the decimal five places left.

10^5 Step 2: The exponent of 10 is 5 since 1.23 times $10^5 = 123{,}000$.

$123{,}000 = 1.23 \times 10^5$ Step 3: Write the original number as the product of the number in Step 1 and the power of 10 in Step 2.

93. $8.11 \times 10^4 = 81{,}100$ Move the decimal point four places to the right.

EXERCISES 3.8

1. $(x - 1)(x - 2) = x(x) - x(2) - 1(x) - 1(-2)$ Multiply. Using FOIL
   $= x^2 - 2x - x + 2$ Simplify.
   $= x^2 - 3x + 2$ Combine like terms.

5. $(z - 5)(z - 3)$
   $= z(z) - z(3) - 5(z) - 5(-3)$ Multiply using FOIL.
   $= z^2 - 3z - 5z + 15$ Simplify.
   $= z^2 - 8z + 15$ Combine like terms.

9. $(x - 7)(x - 5)$
   $= x(x) - x(5) - 7(x) - 7(-5)$ Multiply using FOIL.
   $= x^2 - 5x - 7x + 35$ Simplify.
   $= x^2 - 12x + 35$ Combine like terms.

13. $(a - 12)(a - 6)$
    $= a(a) - a(6) - 12(a) - 12(-6)$ Multiply using FOIL.
    $= a^2 - 6a - 12a + 72$ Simplify.
    $= a^2 - 18a + 72$ Combine like terms.

17. $(2x + 3)(x + 1)$
    $= 2x(x) + 2x(1) + 3x + 3(1)$ Multiply using FOIL.
    $= 2x^2 + 2x + 3x + 3$ Simplify.
    $= 2x^2 + 5x + 3$ Combine like terms.

21. $(4a - 3)(a - 5)$
    $= 4a(a) + 4a(-5) - 3(a) - 3(-5)$ Multiply using FOIL.
    $= 4a^2 - 20a - 3a + 15$ Simplify.
    $= 4a^2 - 23a + 15$ Combine like terms.

25. $(4x + 5)(2x - 3)$
    $= 4x(2x) + 4x(-3) + 5(2x) + 5(-3)$ Multiply using FOIL.
    $= 8x^2 - 12x + 10x - 15$ Simplify.
    $= 8x^2 - 2x - 15$ Combine like terms.

29. $(8 - 5c)(7 - 3c)$
    $= 8(7) + 8(-3c) - 5c(7) - 5c(-3c)$ Multiply using FOIL.
    $= 56 - 24c - 35c + 15c^2$ Simplify.
    $= 15c^2 - 59c + 56$ Combine like terms.

33. $(x - 6)(x + 4) = x^2$
    $x^2 - 2x - 24 = x^2$ Multiply.
    $-2x - 24 = 0$ Subtract x^2 from both sides.
    $-2x = 24$ Add 24 to both sides.
    $x = -12$ Divide both sides by -2.
    $\{-12\}$

37. $(5x - 9)(4x + 3)$
    $= 5x(4x) + 5x(3) - 9(4x) - 9(3)$ Multiply using FOIL.
    $= 20x^2 + 15x - 36x - 27$ Simplify.
    $= 20x^2 - 21x - 27$ Combine like terms.

Exercises 3.8

41. $(5a - 3b)(4a + 2b)$
$= 5a(4a) + 5a(2b) - 3b(4a) - 3b(2b)$ Multiply using FOIL.
$= 20a^2 + 10ab - 12ab - 6b^2$ Simplify.
$= 20a^2 - 2ab - 6b^2$ Combine like terms.

45. $\left(\dfrac{1}{4}x - \dfrac{1}{3}\right)\left(\dfrac{1}{3}x - \dfrac{1}{4}\right)$

$= \left(\dfrac{1}{4}x\right)\left(\dfrac{1}{3}x\right) + \left(\dfrac{1}{4}x\right)\left(-\dfrac{1}{4}\right) - \dfrac{1}{3}\left(\dfrac{1}{3}x\right) - \dfrac{1}{3}\left(-\dfrac{1}{4}\right)$ Multiply using FOIL.

$= \dfrac{1}{12}x^2 - \dfrac{1}{16}x - \dfrac{1}{9}x + \dfrac{1}{12}$ Simplify.

$= \dfrac{1}{12}x^2 - \dfrac{25}{144}x + \dfrac{1}{12}$

$\qquad\qquad\qquad\qquad -\dfrac{1}{16}x - \dfrac{1}{9}x = -\dfrac{9}{144}x - \dfrac{16}{144}x$

$\qquad\qquad\qquad\qquad\qquad\qquad = -\dfrac{25}{144}x.$

49. $(x - 7)(x + 5) = (x + 8)(x + 1)$
$x^2 - 2x - 35 = x^2 + 9x + 8$ Multiply.
$\quad -2x - 35 = 9x + 8$ Subtract x^2 from both sides.
$\qquad\quad -35 = 11x + 8$ Add $2x$ to both sides.
$\qquad\quad -43 = 11x$ Subtract 8 from both sides.

$\qquad -\dfrac{43}{11} = x$ Divide both sides by 11.

$\left\{-\dfrac{43}{11}\right\}$

53. $(3x - 7)(2x + 3) = (6x + 1)(x - 5)$
$6x^2 - 5x - 21 = 6x^2 - 29x - 5$ Multiply.
$\quad -5x - 21 = -29x - 5$ Subtract $6x^2$ from both sides.
$\quad 24x - 21 = -5$ Add $29x$ to both sides.
$\qquad 24x = 16$ Add 21 to both sides.

$\qquad x = \dfrac{2}{3}$ Divide both sides by 24.

$\left\{\dfrac{2}{3}\right\}$

57. number: x

$(x + 2)(x - 5) = x^2 - 4$

$\qquad\qquad\qquad\qquad\qquad\qquad$ $x + 2$: two more than a number.
$\qquad\qquad\qquad\qquad\qquad\qquad$ $x - 5$: five less than the number.
$x^2 - 3x - 10 = x^2 - 4$ $x^2 - 4$: four less than the square of the number.
$\quad -3x - 10 = -4$ Multiply.
$\qquad -3x = 6$ Subtract x^2 from both sides.
$\qquad x = -2$ Add 10 to both sides.
$\qquad\qquad\qquad\qquad\qquad\qquad$ Divide both sides by -3.

The number is -2.

61. number: x

$3x(x - 1) = (x - 8)(3x + 1)$ $3x$: three times a number.
                                             $x - 1$: one less than the number.
                                             $x - 8$: eight less than the number.
                                             $3x + 1$: one more than three times the number.

$3x^2 - 3x = 3x^2 - 23x - 8$ Multiply.
$-3x = -23x - 8$ Subtract $3x^2$ from both sides.
$20x = -8$ Add $23x$ to both sides.

$x = -\dfrac{2}{5}$ Divide both sides by 20.

The number is $-\dfrac{2}{5}$.

65.

	Rate ·	Principle =	Interest
1st part	10% = 0.10	x	$0.10x$
2nd part	12% = 0.12	$12000 - x$	$0.12(12000 - x)$

$\begin{pmatrix}\text{Interest from}\\ \text{10\% part}\end{pmatrix} + \begin{pmatrix}\text{Interest from}\\ \text{12\% part}\end{pmatrix} = \begin{pmatrix}\text{Total}\\ \text{Interest}\end{pmatrix}$

$0.10x + 0.12(12000 - x) = 1332$ Substitute information from the chart.
$0.10x + 1440 - 0.12x = 1332$ Multiply.
$-0.02x + 1440 = 1332$ Combine like terms.
$-0.02x = -108$ Subtract 1440 from both sides.
$x = 5400$ Divide both sides by -0.02.

$x = 5400$
$12000 - x = 12000 - 5400 = 6600$

$5400 at 10%; $6600 at 12%

69. Geri's age now: $2x$ Twice as old as Sam.
    Sam's age now: x

Geri's age in 12 years: $2x + 12$

$2x + 12 = 4x$ Geri's age in 12 years = four times Sam's age now.
$12 = 2x$ Subtract $2x$ from both sides.
$6 = x$ Divide both sides by 2.

$x = 6$
$2x = 2(6) = 12$

Geri's age: 12 years.

73. $(x^4 - 9)(x^4 - 9)$
$= (x^4)(x^4) + (x^4)(-9) - 9(x^4) - 9(-9)$ Multiply using FOIL.
$= x^8 - 9x^4 - 9x^4 + 81$ Simplify.
$= x^8 - 18x^4 + 81$ Combine like terms.

Exercises 3.9

81. $\quad a + \dfrac{x}{b} = -a$

$-a + a + \dfrac{x}{b} = -a - a$ Add $-a$ to both sides.

$\dfrac{x}{b} = -2a$ Simplify.

$b\left(\dfrac{x}{b}\right) = b(-2a)$ Multiply both sides by b.

$x = -2ab$ Simplify.

85. $5y - 7y + 2y - 12y$
$= (5 - 7 + 2 - 12)y$ Distributive property.
$= -12y$ Add and subtract.

EXERCISES 3.9

1. $(x - 8)^2 = x^2 - 2(x)(8) + 8^2$
$= x^2 - 16x + 64$

     Square the first term, subtract twice the product of both terms, and add the square of the last term.

5. $(b + 10)^2 = b^2 + 2(b)(10) + 10^2$
$= b^2 + 20b + 100$

     Square the first term, add twice the product of both terms, and add the square of the last term.

9. $(z + 10)(z - 10) = z^2 - 10^2$
$= z^2 - 100$

     The product of conjugate pairs is the difference of two squares.

13. $(x + y)(x - y) = x^2 - y^2$

     The product of conjugate pairs is the difference of two squares.

17. $(2x + 3)^2 = (2x)^2 + 2(2x)(3) + 3^2$
$= 4x^2 + 12x + 9$

     Square the first term, add twice the product of both terms, and add the square of the last term.

21. $(3a - 4)^2 = (3a)^2 - 2(3a)(4) + 4^2$
$= 9a^2 - 24a + 16$

     Square the first term, subtract twice the product of both terms, and add the square of the last term.

25. $(5y - 9)(5y + 9) = (5y)^2 - 9^2$
$= 25y^2 - 81$

     The product of conjugate pairs is the difference of two squares.

29. $(12 - y)^2 = 12^2 - 2(12)(y) + y^2$
$= 144 - 24y + y^2$
$= y^2 - 24y + 144$

     Square the first term, subtract twice the product of both terms, and add the square of the last term. Rearrange terms.

33. $\quad (x + 3)^2 = x^2 - 3$
$x^2 + 6x + 9 = x^2 - 3$ Multiply.
$6x + 9 = -3$ Subtract x^2 from both sides.
$6x = -12$ Subtract 9 from both sides.
$x = -2$ Divide both sides by 6.

$\{-2\}$

37. $(x + y)^2 = x^2 + 2(x)(y) + y^2$
    $ = x^2 + 2xy + y^2$

Square the first term, add twice the product of both terms, and add the square of the last term.

41. $(3a + 2b)(3a - 2b) = (3a)^2 - (2b)^2$
    $ = 9a^2 - 4b^2$

The product of conjugate pairs is the difference of two squares.

45. $(4c + 3d)^2 = (4c)^2 + 2(4c)(3d) + (3d)^2$
    $ = 16c^2 + 24cd + 9d^2$

Square the first term, add twice the product of both terms, and add the square of the last term.

49. $(x - 8)^2 = (x - 4)(x + 4)$ Multiply.
    $x^2 - 16x + 64 = x^2 - 16$ Subtract x^2 from both sides.
    $-16x + 64 = -16$ Subtract 64 from both sides.
    $-16x = -80$
    $x = 5$ Divide both sides by -16.
    $\{5\}$

53. $(x + 4)^2 = (x + 3)^2 - 5$ Multiply.
    $x^2 + 8x + 16 = x^2 + 6x + 9 - 5$ Combine like terms.
    $x^2 + 8x + 16 = x^2 + 6x + 4$ Subtract x^2 from both sides.
    $8x + 16 = 6x + 4$ Subtract $6x$ from both sides.
    $2x + 16 = 4$ Subtract 16 from both sides.
    $2x = -12$
    $x = -6$ Divide both sides by 2.
    $\{-6\}$

57. $(43)(37)$
    $= (40 + 3)(40 - 3)$ $43 = 40 + 3; \ 37 = 40 - 3$
    $= 40^2 - 3^2$ The product of conjugate pairs is the difference of two squares.
    $= 1600 - 9$
    $= 1591$

61. original side length: x
    original area: $x \cdot x = x^2$

    $(x + 2)^2 = x^2 + 44$ New side length is $x + 2$.
    $x^2 + 4x + 4 = x^2 + 44$ Multiply.
    $4x + 4 = 44$ Subtract x^2 from both sides.
    $4x = 40$ Subtract 4 from both sides.
    $x = 10$ Divide both sides by 4.
    10 ft.

65. $\begin{pmatrix} \text{Area of Original} \\ \text{square} \end{pmatrix} = \begin{pmatrix} \text{Area of new} \\ \text{square} \end{pmatrix} - 656$

    $x^2 = (x + 4)^2 - 656$
    $x^2 = x^2 + 8x + 16 - 656$ Multiply.
    $x^2 = x^2 + 8x - 640$ Combine like terms.
    $0 = 8x - 640$ Subtract x^2 from both sides.
    $640 = 8x$ Add 640 to both sides.
    $80 = x$ Divide both sides by 8.

    80 ft by 80 ft

Exercises 3.10

69. $S = 2\pi r(r + h)$
 $376.8 = 2(3.14)(6)(6 + h)$
 $376.8 = 37.68(6 + h)$
 $10 = 6 + h$
 $4 = h$

Formula.
Replace S with 376.8, π with 3.14 and r with 6.
Multiply.
Divide both sides by 37.68.
Subtract 6 from both sides.

The height is 4 cm.

73. $(a - 2)^4 = (a - 2)^2(a - 2)^2$
$ = (a^2 - 4a + 4)(a^2 - 4a + 4)$

Rewrite as a product.
Square the first term, subtract twice the product of both terms, and add the square of the last term.

$= (a^2 - 4a + 4)(a^2) + (a^2 - 4a + 4)(-4a)$
$ + (a^2 - 4a + 4)(4)$

Distributive property.

$= a^4 - 4a^3 + 4a^2 - 4a^3 + 16a^2 - 16a$
$ + 4a^2 - 16a + 16$

Multiply.

$= a^4 - 8a^3 + 24a^2 - 32a + 16$

Combine like terms.

81. $-\dfrac{5}{3}x^3 - \dfrac{8}{9}x^3 + \dfrac{5}{12}x^3$

$= \left(-\dfrac{5}{3} - \dfrac{8}{9} + \dfrac{5}{12}\right)x^3$

Distributive property.

$= \left(\dfrac{-60}{36} - \dfrac{32}{36} + \dfrac{15}{36}\right)x^3$

36 is the LCD.

$= -\dfrac{77}{36}x^3$

Add and subtract.

85. $21x - 54 = 13x + 19$
 $21x - 54 + 54 = 13x + 19 + 54$
 $21x = 13x + 73$
 $21x - 13x = 13x - 13x + 73$
 $8x = 73$

Add 54 to both sides.

Subtract 13x from both sides.

$\dfrac{8x}{8} = \dfrac{73}{8}$

Divide both sides by 8.

$x = \dfrac{73}{8}$

$\left\{\dfrac{73}{8}\right\}$

EXERCISES 3.10

1. $\dfrac{24a^4}{4a^3} = \dfrac{24}{4} \cdot \dfrac{a^4}{a^3}$

Rewrite as the product of fractions.

$= 6a^{4-3}$
$= 6a$

Divide like bases.

5. $(25y + 60y^2) \div 5y$

   $= \dfrac{25y + 60y^2}{5y}$ 　　　Rewrite as a fraction.

   $= \dfrac{25y}{5y} + \dfrac{60y^2}{5y}$ 　　　Divide each term by $5y$.

   $= 5y^{1-1} + 12y^{2-1}$ 　　　Divide like bases. $y^{1-1} = y^0 = 1$.
   $= 5 + 12y$

9. $\quad\quad\quad x + 3 + \dfrac{2}{x+1}$
   $x+1 \overline{\smash{)}x^2 + 4x + 5}$
   $\quad\quad\underline{x^2 + x}$
   $\quad\quad\quad\quad 3x + 5$
   $\quad\quad\quad\quad\underline{3x + 3}$
   $\quad\quad\quad\quad\quad\quad 2$

   $x^2 \div x = x$. Multiply $x(x+1)$. Subtract. $3x \div x = 3$. Multiply $x(x+1)$. Subtract. The remainder is written over the divisor and added to the rest of the quotient.

13. $\quad\quad\quad x - 2$
    $x-5 \overline{\smash{)}x^2 - 7x + 10}$
    $\quad\quad\underline{x^2 - 5x}$
    $\quad\quad\quad -2x + 10$
    $\quad\quad\quad\underline{-2x + 10}$
    $\quad\quad\quad\quad\quad\quad 0$

    $x^2 \div x = x$. Multiply $x(x-5)$. Subtract. $-2x \div x = -2$. Multiply $-2(x-5)$. Subtract.

17. $\dfrac{-45a^3b^9c^5}{9a^2b^3c^4} = \dfrac{-45}{9} \cdot \dfrac{a^3}{a^2} \cdot \dfrac{b^9}{b^3} \cdot \dfrac{c^5}{c^4}$ 　　　Rewrite as the product of fractions.

    $= -5a^{3-2}b^{9-3}c^{5-4}$ 　　　Divide like bases.
    $= -5ab^6c$

21. $\quad\quad\quad 7x + 8$
    $x-1 \overline{\smash{)}7x^2 + x - 8}$
    $\quad\quad\underline{7x^2 - 7x}$
    $\quad\quad\quad\quad 8x - 8$
    $\quad\quad\quad\quad\underline{8x - 8}$
    $\quad\quad\quad\quad\quad\quad 0$

    $7x^2 \div x = 7x$. Multiply $7x(x-1)$. Subtract. $8x \div x = 8$. Multiply $8(x-1)$. Subtract.

25. $\quad\quad\quad x + 13 + \dfrac{55}{x-5}$
    $x-5 \overline{\smash{)}x^2 + 8x - 10}$
    $\quad\quad\underline{x^2 - 5x}$
    $\quad\quad\quad\quad 13x - 10$
    $\quad\quad\quad\quad\underline{13x - 65}$
    $\quad\quad\quad\quad\quad\quad 55$

    $x^2 \div x = x$. Multiply $x(x-5)$. Subtract. $13x \div x = 13$. Multiply $13(x-5)$. Subtract. The remainder is written over the divisor and added to the rest of the quotient.

Exercises 3.10

29.
$$2b - 1 \overline{)4b^2 - 4b - 5}^{\;2b - 1 - \frac{6}{2b-1}}$$
$$\underline{4b^2 - 2b}$$
$$-2b - 5$$
$$\underline{-2b + 1}$$
$$-6$$

$4b^2 \div 2b = 2b$.
Multiply $2b(2b - 1)$.
Subtract.
$-2b \div 2b = -1$. Multiply $-1(2b - 1)$.
Subtract. The remainder is written over the divisor and added to the rest of the quotient.

33.
$$x - 5 \overline{)2x^2 + 3x - 50}^{\;2x + 13 + \frac{15}{x-5}}$$
$$\underline{2x^2 - 10x}$$
$$13x - 50$$
$$\underline{13x - 65}$$
$$15$$

Write the polynomials with exponents in descending order.

37. $(48a^3b + 64ab^2 - 240a^2b) \div 16ab$

$= \dfrac{48a^3b + 64ab^2 - 240a^2b}{16ab}$

Rewrite as a fraction.

$= \dfrac{48a^3b}{16ab} + \dfrac{64ab^2}{16ab} - \dfrac{240a^2b}{16ab}$

Divide each term by $16ab$.

$= 3a^{3-1}b^{1-1} + 4a^{1-1}b^{2-1} - 15a^{2-1}b^{1-1}$
$= 3a^2 + 4b - 15a$

Divide like bases.
$b^{1-1} = b^0 = 1$. $a^{1-1} = a^0 = 1$.

41.
$$x + 1 \overline{)x^4 + 0x^3 + x^2 + 3x + 1}^{\;x^3 - x^2 + 2x + 1}$$
$$\underline{x^4 + x^3}$$
$$-x^3 + x^2$$
$$\underline{-x^3 - x^2}$$
$$2x^2 + 3x$$
$$\underline{2x^2 + 2x}$$
$$x + 1$$
$$\underline{x + 1}$$

Insert a placeholder for the term containing x^3.

Don't forget to add the opposite to subtract.

45.
$$x - 1 \overline{)x^3 + 0x^2 + 0x - 1}^{\;x^2 + x + 1}$$
$$\underline{x^3 - x^2}$$
$$x^2 + 0x$$
$$\underline{x^2 - x}$$
$$x - 1$$
$$\underline{x - 1}$$

Insert placeholders for the terms containing x^2 and x.

49.
$$
\require{enclose}
\begin{array}{r}
a^4 + a^3b + a^2b^2 + ab^3 + b^4 \\
a-b \enclose{longdiv}{a^5 + 0a^4b + 0a^3b^2 + 0a^2b^3 + 0ab^4 - b^5} \\
\underline{a^5 - a^4b} \\
a^4b + 0a^3b^2 \\
\underline{a^4b - a^3b^2} \\
a^3b^2 + 0a^2b^3 \\
\underline{a^3b^2 - a^2b^3} \\
a^2b^3 + 0ab^4 \\
\underline{a^2b^3 - ab^4} \\
ab^4 - b^5 \\
\underline{ab^4 - b^5} \\
0
\end{array}
$$

Insert placeholders for the terms containing a^4b, a^3b^2, a^2b^3 and ab^4.

53.
$$
\begin{array}{r}
x^2 - 4x + 3 \\
x^2 + 4x + 3 \enclose{longdiv}{x^4 + 0x^3 - 10x^2 + 0x + 9} \\
\underline{x^4 + 4x^3 + 3x^2} \\
-4x^3 - 13x^2 + 0x \\
\underline{-4x^3 - 16x^2 - 12x} \\
3x^2 + 12x + 9 \\
\underline{3x^2 + 12x + 9} \\
0
\end{array}
$$

Insert placeholders for the terms containing x^3 and x.

Don't forget to add the opposite to subtract.

57.
$$
\begin{array}{r}
4x + 5 \\
3x - 2 \enclose{longdiv}{12x^2 + 7x - 10} \\
\underline{12x^2 - 8x} \\
15x - 10 \\
\underline{15x - 10} \\
0
\end{array}
$$

$\dfrac{A}{w} = l.$

$4x + 5$ ft

61.
$$
\begin{array}{r}
3x + 7 \\
2x - 5 \enclose{longdiv}{6x^2 - x - n} \\
\underline{6x^2 - 15x} \\
14x - n \\
\underline{14x - 35} \\
-n + 35
\end{array}
$$

$6x^2 \div 2x = 3.$
Multiply $3x(2x - 5)$.
Subtract.
$14x \div 2x = 7$. Multiply $7(2x - 5)$.
Subtract.

Remainder $= 0$
$-n + 35 = 0$
$35 = n$

Solve for n.

65.
$V = lwh$
$6x^3 + 23x^2 - 6x - 8 = l(x + 4)(2x + 1)$
$6x^3 + 23x^2 - 6x - 8 = l(2x^2 + 9x + 4)$

Formula.
Substitute.
Multiply.

$$\begin{array}{r}3x - 2\\2x^2 + 9x + 4\overline{\smash{\big)}6x^3 + 23x^2 - 6x - 8}\\\underline{6x^3 + 27x^2 + 12x}\\-4x^2 - 18x - 8\\\underline{-4x^2 - 18x - 8}\\0\end{array}$$

Divide.

length: $3x - 2$ in.

69. $\dfrac{4(x-3)^3 - 2(x-3)^2 + (x-3)}{(x-3)}$

$= \dfrac{4(x-3)^3}{(x-3)} - \dfrac{2(x-3)^2}{(x-3)} + \dfrac{(x-3)}{(x-3)}$ Divide each term by $x - 3$.

$= 4(x-3)^{3-1} - 2(x-3)^{2-1} + (x-3)^{1-1}$ Divide like bases.
$= 4(x-3)^2 - 2(x-3) + 1$ $(x-3)^{1-1} = (x-3)^0 = 1$.
$= 4(x^2 - 6x + 9) - 2x + 6 + 1$ Multiply.
$= 4x^2 - 24x + 36 - 2x + 7$
$= 4x^2 - 26x + 43$ Combine like terms.

77. $\dfrac{a^7 b^{12} c^8}{a^6 b^6 c^5} = a^{7-6} b^{12-6} c^{8-5}$ Subtract the exponents.

$= ab^6 c^3$

81.

	Cost of ticket	·	Number of tickets	=	Total Sales
$15 tickets	15		n		$15n$
$23 tickets	23		n		$23n$

$\begin{pmatrix}\text{Sales from}\\ \$15\text{ tickets}\end{pmatrix} + \begin{pmatrix}\text{Sales from}\\ \$23\text{ tickets}\end{pmatrix} = \begin{pmatrix}\text{Total}\\ \text{Sales}\end{pmatrix}$

$15n + 23n = 8550$ Substitute from the chart.
$38n = 8550$ Combine like terms.
$n = 225$ Divide both sides by 38.

225 of each kind of ticket.

CHAPTER 3 CONCEPT REVIEW

1. False; $x^2 \cdot y^3$ cannot be simplified further.

2. True

3. True

4. False; $(-3)^{-2} = \dfrac{1}{(-3)^2} = \dfrac{1}{9}$.

5. True

6. False; $623.5 = 6.235 \times 10^2$

7. False; $x^2 + \dfrac{1}{x}$ is not a polynomial.

8. True

9. False; A polynomial can contain one term.

10. False; A binomial is a polynomial with two terms.

11. True

12. True

13. True

14. False; Like terms are terms containing the same variables raised to the same powers.

15. False; $14x + 31x = 45x$.

16. True

17. True

18. True

19. True

20. False; The FOIL system is a shortcut for multiplying two binomials.

21. False; $(3x + 2)(2x - 3) = 6x^2 - 5x - 6$.

22. False; Conjugates are the sum and difference of the same two terms.

23. True

24. True

Chapter 3 Test   91

25. False;

$$\begin{array}{r}x^3 + x^2y + xy^2 + y^3\\ x-y\overline{\smash{)}x^4 + 0x^3y + 0x^2y^2 + 0xy^3 - y^4}\\ \underline{x^4 - x^3y}\\ x^3y + 0x^2y^2 + 0xy^3 - y^4\\ \underline{x^3y - x^2y^2}\\ x^2y^2 + 0xy^3 - y^4\\ \underline{x^2y^2 - xy^3}\\ xy^3 - y^4\\ \underline{xy^3 - y^4}\\ 0\end{array}$$

26. False; $(x - 1)(x^2 + x + 1) = x^3 - 1.$

CHAPTER 3 TEST

1. $-5x + 3x + 11x = (-5 + 3 + 11)x$ Distributive property.
    $ = 9x$ Simplify.

2. $(2x + 9)(2x - 9) = (2x)^2 - 9^2$ The product of conjugate pairs is the difference
    $ = 4x^2 - 81$ of two squares.

3. $\begin{array}{r}x - 7\\ x - 8\overline{\smash{)}x^2 - 15x + 56}\\ \underline{x^2 - 8x}\\ -7x + 56\\ \underline{-7x + 56}\\ 0\end{array}$ $x^2 \div x = x.$
   Multiply $x(x - 8)$.

   Subtract.
   $-7x \div x = -7.$ Multiply $-7(x - 8)$.

4. $(12a - 3b + 4) + (8a - 9b - 12)$
   $= (12a + 8a) + (-3b - 9b) + (4 - 12)$ Group like terms.
   $= (12a + 8a) + [-3b + (-9b)] + [4 + (-12)]$ Change to addition.
   $= 20a - 12b - 8$ Combine like terms.

5. Trinomial   Three terms.

6. $(3m - 11) - (5m + 6) = 20$
   $(3m - 11) + (-5m - 6) = 20$ Change to addition.
   $-2m - 17 = 20$ Combine like terms.
   $-2m = 37$ Add 17 to both sides.

   $m = \dfrac{-37}{2}$ Divide both sides by -2.

   $\left\{-\dfrac{37}{2}\right\}$

7. $(2x - 7)(x + 6) = 2x(x) + 2x(6) - 7(x) - 7(6)$ Multiply using FOIL.
   $ = 2x^2 + 12x - 7x - 42$ Simplify.
   $ = 2x^2 + 5x - 42$ Combine like terms.

8. $-8x^9 + 5x^5 - 7x^3 - 3x^2 + 4x + 6$ Write with exponents decreasing as you read from
    left to right.

9. $6x - 5 = 8x - 7$
   $-5 = 2x - 7$ Subtract $6x$ from both sides.
   $2 = 2x$ Add 7 to both sides.
   $1 = x$ Divide both sides by 2.

   $\{1\}$

10. $-4xyz(-12x - 7y + 2z)$
    $= -4xyz(-12x) - 4xyz(-7y) - 4xyz(2z)$ Distributive property.
    $= 48x^2yz + 28xy^2z - 8xyz^2$ Multiply.

11. $4x - 5y + 8$
    $\underline{6x + 2y - 7}$

    $4x - 5y + 8$
    $\underline{-6x - 2y + 7}$ Subtract by adding the opposite.
    $-2x - 7y + 15$

12. $(2x - 3y) + (4x + 6y) + (2y - 5x)$
    $= (2x + 4x - 5x) + (-3y + 6y + 2y)$ Group like terms.
    $= x + 5y$ Combine like terms.

13. $(m - 10)^2 = m^2 - 2(m)(10) + 10^2$ Square the first term, subtract twice the product
    $ = m^2 - 20m + 100$ of both terms, and add the square of the last term.

14. $(2a - 3b)(4a^2 + 3ab - 2b^2)$
    $= (2a - 3b)(4a^2) (2a - 3b)(3ab)$ Distributive property.
    $+ (2a - 3b)(-2b^2)$
    $= 8a^3 - 12a^2b + 6a^2b - 9ab^2 - 4ab^2 + 6b^3$ Multiply.
    $= 8a^3 - 6a^2b - 13ab^2 + 6b^3$ Combine like terms.

15. $(ay)^5 = a^5y^5$ The exponent 5 applies to both factors inside the parentheses.

16. $(x - 3)(x + 4) = x^2 + 4$
    $x^2 + x - 12 = x^2 + 4$ Multiply.
    $x - 12 = 4$ Subtract x^2 from both sides.
    $x = 16$ Add 12 to both sides.

    $\{16\}$

17. $m^9 \cdot m^5 \cdot m^7 = m^{9+5+7}$ Add the exponents.
    $ = m^{21}$

18. $3.6 \times 10^{-8} = 0.000000036$ Move the decimal point 8 places to the left.

19. $(y - 5)(y^2 + 3y + 2)$
    $= (y - 5)(y^2) + (y - 5)(3y) + (y - 5)(2)$ Distributive property.
    $= y^3 - 5y^2 + 3y^2 - 15y + 2y - 10$ Multiply.
    $= y^3 - 2y^2 - 13y - 10$ Combine like terms.

Chapter 3 Test

20. $12x + 3 - 8x = -9 + 7x + 6$
    $4x + 3 = -3 + 7x$ Combine like terms.
    $3 = -3 + 3x$ Subtract $4x$ from both sides.
    $6 = 3x$ Add 3 to both sides.
    $2 = x$ Divide both sides by 3.
    {2}

21. $m^{-6}n^3 = \dfrac{n^3}{m^6}$ The definition of a negative exponent.

22. $(4x - 9)(9x + 4)$
    $= (4x)(9x) + (4x)(4) - 9(9x) - 9(4)$ Multiply using FOIL.
    $= 36x^2 + 16x - 81x - 36$ Simplify.
    $= 36x^2 - 65x - 36$ Combine like terms.

23. $4a(a^2 - 3a - 2)$
    $= 4a(a^2) - 4a(3a) - 4a(2)$ Distributive property.
    $= 4a^3 - 12a^2 - 8a$ Multiply.

24. $(14x^2 - 5x) - (6x - 5)$
    $= (14x^2 - 5x) + (-6x + 5)$ Change to addition.
    $= 14x^2 + (-5x - 6x) + 5$ Group like terms.
    $= 14x^2 - 11x + 5$ Combine like terms.

25. $(x - 3)(x + 3) = (x - 2)(x - 5) + 2$
    $x^2 - 9 = x^2 - 7x + 10 + 2$ Multiply.
    $x^2 - 9 = x^2 - 7x + 12$ Combine like terms.
    $-9 = -7x + 12$ Subtract x^2 from both sides.
    $-21 = -7x$ Subtract 12 from both sides.
    $3 = x$ Divide both sides by -7.
    {3}

26.
$$\require{enclose}\begin{array}{r}x^2 - x - 2 + \dfrac{8}{x+1}\\[2pt] x+1\enclose{longdiv}{x^3 + 0x^2 - 3x + 6}\\[2pt] \underline{x^3 + x^2}\\[2pt] -x^2 - 3x + 6\\[2pt] \underline{-x^2 - x}\\[2pt] -2x + 6\\[2pt] \underline{-2x - 2}\\[2pt] 8\end{array}$$

$x^3 + x = x^2$.
Multiply $x^2(x + 1)$.
Subtract.
$-x^2 + x = -x$. Multiply $-x(x + 1)$.
Subtract.
$-2x + x = -2$. Multiply $-2(x + 1)$.
Subtract. The remainder is written over the divisor and added to the rest of the quotient.

27. $12x(x - 5) = (3x - 20)(4x + 3)$
    $12x^2 - 60x = 12x^2 - 71x - 60$ Multiply.
    $-60x = -71x - 60$ Subtract $12x^2$ from both sides.
    $11x = -60$ Add $71x$ to both sides.

$x = -\dfrac{60}{11}$ Divide both sides by 11.

$\left\{-\dfrac{60}{11}\right\}$

28. $\dfrac{4a^5}{2a^2} = 2a^{5-2}$ Subtract the exponents.

    $= 2a^3$

29. $(12abc - 16a^2bc^2 + 24abc^2) \div 4abc$

    $= \dfrac{12abc - 16a^2bc^2 + 24abc^2}{4abc}$ Rewrite as a fraction.

    $= \dfrac{12abc}{4abc} - \dfrac{16a^2bc^2}{4abc} + \dfrac{24abc^2}{4abc}$ Divide each term by $4abc$.

    $= 3a^{1-1}b^{1-1}c^{1-1} - 4a^{2-1}b^{1-1}c^{2-1} + 6a^{1-1}b^{1-1}c^{2-1}$ Divide like bases.
    $= 3 - 4ac + 6c$ $a^{1-1} = a^0 = 1.$ $b^{1-1} = b^0 = 1.$ $c^{1-1} = c^0 = 1.$

30. number: x

    $3x - (5x - 6) = 12$ $3x$: Three times a number.
    $5x - 6$: Six less than five times a number.

    $3x - 5x + 6 = 12$ Change to addition.
    $-2x + 6 = 12$ Combine like terms.
    $-2x = 6$ Subtract 6 from both sides.
    $x = -3$ Divide both sides by -2.

    The number is -3.

31.

	Rate	·	Time	=	Distance
Traveler	7000		t		$7000t$
Voyager	8600		t		$8600t$

$\begin{pmatrix}\text{Distance traveled}\\\text{by }Voyager\end{pmatrix} - \begin{pmatrix}\text{Distance traveled}\\\text{by }Traveler\end{pmatrix} = \begin{pmatrix}\text{Distance between}\\\text{the two}\end{pmatrix}$

$8600t - 7000t = 5600$ Substitute from the chart.
$1600t = 5600$ Combine like terms.

$t = 3\dfrac{1}{2}$ Divide both sides by 1600.

$3\dfrac{1}{2}$ hours

CHAPTER 4

EXERCISES 4.1

1. $(x - 7)(x - 1) = x^2 - x - 7x + 7$
   $ = x^2 - 8x + 7$ Multiply using FOIL.
   Yes

5. $(w + 3)(w - 6) = w^2 - 6w + 3w - 18$
   $ = w^2 - 3w - 18$ Multiply using FOIL.
   Yes

9. $(y - 8)(y + 8) = 0$
   $y - 8 = 0$ or $y + 8 = 0$ Set each factor equal to 0 using the zero-product property.
   $y = 8$ or $y = -8$ Solve.
   $\{-8, 8\}$

13. $(t - 15)(t - 20) = 0$
    $t - 15 = 0$ or $t - 20 = 0$ Set each factor equal to 0.
    $t = 15$ or $t = 20$ Solve.
    $\{15, 20\}$

17. $z(z - 21) = z^2 - 21z$ Distributive property.
    Yes

21. $(y - 16)(y + 16) = y^2 - 16^2$ Multiply conjugates.
    $ = y^2 - 256$ Doesn't equal the given polynomial.
    No

25. $(y + 3.4)(y - 0.6) = 0$
    $y + 3.4 = 0$ or $y - 0.6 = 0$ Set each factor equal to 0.
    $y = -3.4$ or $y = 0.6$ Solve.
    $\{-3.4, 0.6\}$

29. $\left(v + \dfrac{1}{3}\right)\left(v - \dfrac{2}{5}\right) = 0$

    $v + \dfrac{1}{3} = 0$ or $v - \dfrac{2}{5} = 0$ Set each factor equal to 0.

    $v = -\dfrac{1}{3}$ or $v = \dfrac{2}{5}$ Solve.

    $\left\{-\dfrac{1}{3}, \dfrac{2}{5}\right\}$

33. $(2s - 5)(s - 6) = 0$
    $2s - 5 = 0$ or $s - 6 = 0$ Set each factor equal to 0.
    $2s = 5$ or $s = 6$

$s = \dfrac{5}{2}$ Solve.

$\left\{\dfrac{5}{2}, 6\right\}$

37. $(5w - 11)(5w - 11) = 0$

     $5w - 11 = 0$ or $5w - 11 = 0$
          $5w = 11$ or $5w = 11$

            $w = \dfrac{11}{5}$ or $w = \dfrac{11}{5}$ A double root.

     $\left\{\dfrac{11}{5}\right\}$

41. $(7r + 6)(2r + 13) = 0$
     $7r + 6 = 0$ or $2r + 13 = 0$ Set each factor equal to 0.
          $7r = -6$ or $2r = -13$ Solve.

            $r = -\dfrac{6}{7}$ or $r = -\dfrac{13}{2}$

     $\left\{-\dfrac{13}{2}, -\dfrac{6}{7}\right\}$

45. $(14x + 91)(3x - 5) = 0$
     $14x + 91 = 0$ or $3x - 5 = 0$ Set each factor equal to 0.
            $14x = -91$ or $3x = 5$ Solve.

               $x = -\dfrac{13}{2}$ or $x = \dfrac{5}{3}$

     $\left\{\dfrac{5}{3}, -\dfrac{13}{2}\right\}$

49. $16t(5 - t) = h$
     $16t(5 - t) = 0$ Formula.
     $16t = 0$ or $5 - t = 0$ Replace h with 0.
       $t = 0$ or $5 = t$ Set each factor equal to 0.
                                         Solve.

     0 sec, 5 sec

53. 1st number: $x + 13$
     2nd number: x
     $x(x + 13) = 0$
         $x = 0$ or $x + 13 = 0$ Product is 0.
                               $x = -13$ Set each factor equal to 0.
     $x + 13 = 13$ $x + 13 = 0$ Solve.

     The numbers are 0 and 13 or 0 and -13.

57. One possible answer: If when multiplying factors the resulting product is zero, then at least one of those factors must equal zero.

Exercises 4.2

61. $\left(2x\left(\dfrac{3}{5} - \dfrac{3}{4}x\right)\left(\dfrac{6}{7} + \dfrac{5}{3}x\right) = 0\right.$

$2x = 0$ or $\dfrac{3}{5} - \dfrac{3}{4}x = 0$ or $\dfrac{6}{7} + \dfrac{5}{3}x = 0$ Zero-product property.

$x = 0$ $\dfrac{3}{5} = \dfrac{3}{4}x$ $\dfrac{5}{3}x = -\dfrac{6}{7}$ Solve.

               $\dfrac{4}{5} = x$ $x = -\dfrac{18}{35}$

$\left\{-\dfrac{18}{35}, 0, \dfrac{4}{5}\right\}$

65. $23 - 8x - 24x = -15x + 34$

        $23 - 32x = -15x + 34$ Combine like terms.

            $-32x = -15x + 11$ Subtract 23 from both sides.

               $-17x = 11$ Add $15x$ to both sides.

                  $x = -\dfrac{11}{17}$ Divide both sides by -17.

$\left\{-\dfrac{11}{17}\right\}$

69. $4x + 2 < 2x - 8$

        $2x + 2 < -8$ Subtract $2x$ from both sides.

            $2x < -10$ Subtract 2 from both sides.

              $x < -5$ Divide both sides by 2.

$\{x \mid x < -5\}$

73. $6x + 3 - 5x > 5x + 6 - 7x$

         $x + 3 > -2x + 6$ Combine like terms.

           $3x + 3 > 6$ Add $2x$ to both sides.

              $3x > 3$ Subtract 3 from both sides.

               $x > 1$ Divide both sides by 3.

$\{x \mid x > 1\}$

EXERCISES 4.2

1. $7x - 7w = 7(x) - 7(w)$ 7 is the GCF.

            $= 7(x - w)$ 7 is factored out.

5. $2ab + 2ac = 2a(b) + 2a(c)$ $2a$ is the GCF.

               $= 2a(b + c)$ $2a$ is factored out.

9. $xy + yz = x(y) + x(z)$ x is the GCF.

            $= x(y + z)$ x is factored out.

13. $4ab - 10ac = 2a(2b) - 2a(5c)$ $2a$ is the GCF.

                $= 2a(2b - 5c)$ $2a$ is factored out.

17. $15xyz + 7wxz = xz(15y) + xz(7w)$ xz is the GCF.

                 $= xz(15y + 7w)$ xz is factored out.

21.   Prime

25. $4ab + 8bc - 6cd = 2(2ab) + 2(4bc) + 2(-3cd)$ 2 is the GCF.
$= 2(2ab + 4bc - 3cd)$ 2 is factored out.

29. $24ab - 28ac + 42ad$ $2a$ is the GCF.
$= 2a(12b) + 2a(-14c) + 2a(21d)$ $2a$ is factored out.
$= 2a(12b - 14c + 21d)$

33. $6x^2 - 6x = 0$
$6x(x - 1) = 0$ Factor the left side. The GCF is $6x$.
$6x = 0$ or $x - 1 = 0$ Zero-product property.
$x = 0$ or $x = 1$ Solve.

{0, 1}

37. $-8a^2 + 24a = 0$
$-8a(a - 3) = 0$ Factor the left side. The GCF is $-8a$.
$-8a = 0$ or $a - 3 = 0$ Zero-product property.
$a = 0$ or $a = 3$ Solve.

{0, 3}

41. $x^2y^2 - x^3y^3 + x^4y^4$
$= x^2y^2(1) + x^2y^2(-xy) + x^2y^2(x^2y^2)$ x^2y^2 is the GCF.
$= x^2y^2(1 - xy + x^2y^2)$ x^2y^2 is factored out.

45. $3m^2n^2 - 6mn - 12m^2n - 15mn^2$
$= 3mn(mn) + 3mn(-2) + 3mn(-4m) + 3mn(-5n)$ $3mn$ is the GCF.
$= 3mn(mn - 2 - 4m - 5n)$ $9x^2y^2$ is factored out.

Exercises 4.2

49. $9x^2y^2 - 72x^3y^3 - 567x^4y^4$
    $= 9x^2y^2(1) + 9x^2y^2(-8xy) + 9x^2y^2(-63x^2y^2)$ $9x^2y^2$ is the GCF.
    $= 9x^2y^2(1 - 8xy - 63x^2y^2)$ $9x^2y^2$ is factored out.

53. $9m^2n^2 + 18mn - 36m^2n - 45mn^2$
    $= 9mn(mn) + 9mn(2) + 9mn(-4m) + 9mn(-5n)$ $9mn$ is the GCF.
    $= 9mn(mn + 2 - 4m - 5n)$ $9mn$ is factored out.

57. $35y^2 - 7y = 0$
    $7y(5y - 1) = 0$ Factor the left side. The GCF is $7y$.
    $7y = 0$ or $5y - 1 = 0$ Zero-product property.
    $y = 0$ or $5y = 1$ Solve.
    $\qquad\qquad\qquad y = \dfrac{1}{5}$

    $\left\{0, \dfrac{1}{5}\right\}$

61. $\qquad t^2 = -17t$
    $t^2 + 17t = 0$ Add $17t$ to both sides.
    $t(t + 17) = 0$ Factor the left side. The GCF is t.
    $t = 0$ or $t + 17 = 0$ Zero-product property.
    $\qquad\qquad t = -17$ Solve.

    $\{-17, 0\}$

65. Pete's age: x
    Thelma's age: $2x$

    $x(2x) = 6x$ The product of their ages is 6 times Pete's age.
    $2x^2 = 6x$
    $2x^2 - 6x = 0$ Subtract $6x$ from both sides.
    $2x(x - 3) = 0$ Factor the left side. The GCF is $2x$.
    $2x = 0$ or $x - 3 = 0$ Zero-product property.
    $x = 0$ or $\qquad x = 3$ Solve.
    $\qquad\qquad 2x = 2(3) = 6$

    Pete's age: 3 yrs;   Thelma's age: 6 yrs

69. radius: r

    $\pi r^2 = 2\pi r$ Area = πr^2. Circumference = $2\pi r$.
    $\pi r^2 - 2\pi r = 0$ Subtract $2\pi r$ from both sides.
    $\pi r(r - 2) = 0$ Factor the left side. πr is the GCF.

$\pi r = 0$ or $r - 2 = 0$ Zero-product property.
$r = 0$ or $r = 2$ Solve.

Since $r > 0$, $r = 2$ m.

73. One possible answer: The greatest common factor is the product of the largest integral factor that is common to the coefficients and the variable factor(s) that is (are) common to each term using the smallest exponent of the common variable.

77. $(3a + 4b)(2a - 5b) + (3a + 4b)(3a + 3b)$
 $= (3a + 4b)[(2a - 5b) + (3a + 3b)]$ The GCF is $3a + 4b$.
 $= (3a + 4b)(5a - 2b)$ Combine like terms in the brackets.

85. $3a^2b(b + a) - 2c^2(a + b)$
 $= 3a^2b^2 + 3a^3b - 2ac^2 - 2bc^2$ Multiply.

89. $1.57 - 2.1(-3.4x) + 8 \geq 16 - 5.2(1.1x)$
 $1.57 + 7.14x + 8 \geq 16 - 5.72x$ Multiply.
 $7.14x + 9.57 \geq 16 - 5.72x$ Combine like terms.
 $12.86x + 9.57 \geq 16$ Add $5.72x$ to both sides.
 $12.86x \geq 6.43$ Subtract 9.57 from both sides.
 $x \geq 0.5$ Divide both sides by 12.86.
 $\{x \mid x \geq 0.5\}$

EXERCISES 4.3

1. $cx + cy + 3x + 3y$
 $= (cx + cy) + (3x + 3y)$ Group the first two and the last two terms.
 $= (x + y) \cdot c + (x + y) \cdot 3$ The GCF of the first group is c and of the second group is 3. Factor these out.
 $= (x + y)(c + 3)$ Factor out the common polynomial $(x + y)$ from each term.

5. $x^2 + x + xw + w$
 $= (x^2 + x) + (xw + w)$ Group the first two and the last two terms.
 $= (x + 1) \cdot x + (x + 1) \cdot w$ Factor x from the first group and w from the second group.
 $= (x + 1)(x + w)$ Factor out $x + 1$.

9. $x^2 + 2x + 5x + 10 = 0$
 $(x^2 + 2x) + (5x + 10) = 0$
 $x(x + 2) + 5(x + 2) = 0$ Factor the left side by grouping.
 $(x + 2)(x + 5) = 0$
 $x + 2 = 0$ or $x + 5 = 0$ Zero-product property.
 $x = -2$ or $x = -5$
 $\{-5, -2\}$

13. $y^2 + 7y + y + 7 = 0$
 $(y^2 + 7y) + (y + 7) = 0$
 $y(y + 7) + 1(y + 7) = 0$ Factor the left side by grouping.
 $(y + 7)(y + 1) = 0$

Exercises 4.3

$$y + 7 = 0 \quad \text{or} \quad y + 1 = 0$$
$$y = -7 \quad \text{or} \quad y = -1$$

Zero-product property.

$\{-7, -1\}$

17. $x^5 + x^3 + x^2 + 1$
$(x^5 + x^3) + (x^2 + 1)$ Group the first two and the last two terms.
$(x^2 + 1) \cdot x^3 + (x^2 + 1) \cdot 1$ Factor x^3 from the first group and 1 from the second group.
$(x^2 + 1)(x^3 + 1)$ Factor out $x^2 + 1$.

21. $ax + a + b + bx$
$= (ax + a) + (b + bx)$ Group the first two and the last two terms.
$= (x + 1) \cdot a + (1 + x) \cdot b$ Factor a from the first group and b from the second group.
$= (x + 1)(a + b)$ Factor out $x + 1$. Remember $x + 1 = 1 + x$.

25. $24a^2b - 8a + 15abc - 5c$
$= (24a^2b - 8a) + (15abc - 5c)$ Group the first two and the last two terms.
$= (3ab - 1) \cdot 8a + (3ab - 1) \cdot 5c$ Factor $8a$ from the first group and $5c$ from the second group.
$= (3ab - 1)(8a + 5c)$ Factor out $3ab - 1$.

29. $3y^2 - 24y + 2y - 16 = 0$
$3y(y - 8) + 2(y - 8) = 0$
$(y - 8)(3y + 2) = 0$ Factor the left side by grouping.

$y - 8 = 0 \quad \text{or} \quad 3y + 2 = 0$ Zero-product property.
$y = 8 \quad\quad\quad\quad 3y = -2$
$\quad\quad\quad\quad\quad\quad\quad y = -\dfrac{2}{3}$

$\left\{-\dfrac{2}{3}, 8\right\}$

33. $\quad\quad\quad\quad\quad 6t^2 - 12t = t - 2$
$\quad\quad\quad\quad 6t^2 - 12t - t + 2 = 0$ Subtract t from both sides.
$\quad\quad\quad\quad 6t(t - 2) - 1(t - 2) = 0$ Add 2 to both sides.
$\quad\quad (t - 2)(6t - 2) - 1(t - 2) = 0$ Factor the left side by grouping.
$t - 2 = 0 \quad \text{or} \quad 6t - 1 = 0$ Zero-product property.
$t = 2 \quad \text{or} \quad\quad 6t = 1$
$\quad\quad\quad\quad\quad\quad\quad\quad t = \dfrac{1}{6}$

$\left\{\dfrac{1}{6}, 2\right\}$

37. $9x + 6 - 6ax - 4a$
$= (9x + 6) + (-6ax - 4a)$ Group the first two and the last two terms.
$= 3(3x + 2) - 2a(3x + 2)$ Factor 3 from the first group and $-2a$ from the second group.
$= (3x + 2)(3 - 2a)$ Factor out $3x + 2$.

41. $12a^2b^2c^2 + 4bc + 9a^2bc + 3$
$= (12a^2b^2c^2 + 4bc) + (9a^2bc + 3)$ Group the first two and the last two terms.
$= 4bc(3a^2bc + 1) + 3(3a^2bc + 1)$ Factor $4bc$ from the first group and 3 from the second group.
$= (3a^2bc + 1)(4bc + 3)$ Factor out $3a^2bc + 1$.

45. $a^3 + 2a^2 + a^2b + 2ab + ab^2 + 2b^2$
    $= a^3 + a^2b + ab^2 + 2a^2 + 2ab + 2b^2$
    $= (a^3 + a^2b + ab^2) + (2a^2 + 2ab + 2b^2)$
    $= a(a^2b + ab + b^2) + 2(a^2 + ab + b^2)$
    $= (a^2 + ab + b^2)(a + 2)$

    Rearrange the terms.
    Group the first three terms and the last three terms.
    Factor a from the first group and 2 from the second group.
    Factor out $a^2 + ab + b^2$.

49. $\quad 6y^2 + 9y + 8y + 12 = 0$
    $\quad 3y(2y + 3) - + 4(2y + 3) = 0$
    $\qquad (2y + 3)(3y + 4) = 0$
    $2y + 3 = 0 \quad$ or $\quad 3y + 4 = 0$
    $\quad 2y = -3 \quad$ or $\quad 3y = -4$
    $\quad 2y = -\dfrac{3}{2} \quad$ or $\quad y = -\dfrac{4}{3}$

    $\left\{-\dfrac{3}{2}, -\dfrac{4}{3}\right\}$

    Factor the left side by grouping.

    Zero-product property.

53. $\qquad 15t^2 - 20t = 28 - 21t$
    $\quad 15t^2 - 20t + 21t - 28 = 0$
    $\quad 5t(3t - 4) + 7(3t - 4) = 0$
    $\qquad (3t - 4)(5t + 7) = 0$
    $3t - 4 = 0 \quad$ or $\quad 5t + 7 = 0$
    $\quad 3t = 4 \quad$ or $\quad 5t = -7$
    $\quad t = \dfrac{4}{3} \qquad \quad t = -\dfrac{7}{5}$

    $\left\{-\dfrac{7}{5}, \dfrac{4}{3}\right\}$

    Add $21t$ to both sides and subtract 28 from both sides.
    Factor the left side by grouping.

    Zero-product property.

57. One possible answer: A four-term polynomial can be grouped three different ways. The first term can be grouped with the second term or the third term or the fourth term.

61. $8acd + 4bcd - 16ace - 8bce$
    $= 4c(2ad + bd - 4ae - 2be)$
    $= 4c[(2ad + bd) + (-4ae - 2be)]$

    $= 4c[d(2a + b) - 2e(2a + b)]$
    $= 4c(2a + b)(d - 2e)$

    Factor out $4c$.
    Inside the parentheses group the first two terms and the last two terms.
    Factor d from the first group and $-2e$ from the second group.
    Factor out $2a + b$.

65. $(c - 5)(c - 12) = c^2 - 12c - 5c + 60$
    $\qquad\qquad\qquad = c^2 - 17c + 60$

    Multiply by using FOIL.
    Combine like terms.

69. $(a + 5)(a + 11) = a^2 + 11a + 5a + 55$
    $\qquad\qquad\qquad = a^2 + 16a + 55$

    Multiply by using FOIL.
    Combine like terms.

73. $(15a - 16) - (21 - 5a) = 12a + 34)$
    $\qquad\qquad\qquad\qquad\quad + (3a - 1)$
    $(15a - 16) + (-21 + 5a) = (12a + 34)$
    $\qquad\qquad\qquad\qquad\quad + (3a - 1)$

    Change to addition.

Exercises 4.4

$$20a - 37 = 15a + 33$$ Combine like terms.
$$5a - 37 = 33$$ Subtract $15a$ from both sides.
$$5a = 70$$ Add 37 to both sides.
$$a = 14$$ Divide both sides by 5.
$$\{14\}$$

EXERCISES 4.4

1. Product: Sum:
   $mn = 8$ $m + n = 6$
   $(1)(8)$ $1 + 8 = 9$ List the pairs of positive integers whose product is 8
   $(2)(4)$ $2 + 4 = 6$ and find the sum.
   $(2)(4)$ $2 + 4 = 6$ The correct sum.

   $x^2 + 6x + 8 = (x + 2)(x + 4)$

5. Product: Sum:
   $mn = -27$ $m + n = 6$
   $(1)(-27)$ $1 + (-27) = -26$ List the pairs whose product is -27 and find the sum.
   $(-1)(27)$ $-1 + 27 = 26$
   $(3)(-9)$ $3 + (-9) = -6$
   $(-3)(9)$ $-3 + 9 = 6$
   $(-3)(9)$ $-3 + 9 = 6$ The correct sum.

   $y^2 + 6y - 27 = (y - 3)(y + 9)$

9. $x^2 + 16x + 15 = 0$
   $(x + 15)(x + 1) = 0$ Factor the left side.
   $x + 15 = 0$ or $x + 1 = 0$ Zero-product property.
        $x = -15$ or $x = -1$

   $\{-15, -1\}$

13. $w^2 - 11w - 26 = 0$
    $(w - 13)(w + 2) = 0$ Factor the left side.
    $w - 13 = 0$ or $w + 2 = 0$ Zero-product property.
         $w = 13$ or $w = -2$

    $\{-2, 13\}$

17. Product: Sum:
    $mn = 21$ $m + n = 22$
    $(1)(21) = 21$ $1 + 21 = 22$ List the pairs of positive integers whose product is 21
    $(3)(7) = 21$ $3 + 7 = 10$ and find the sum.
    $(1)(21) = 21$ $1 + 21 = 22$ The correct sum.

    $x^2 + 22x + 21 = (x + 1)(x + 21)$

21. Product: Sum:
    $mn = -33$ $m + n = -8$
    $(1)(-33) = -33$ $1 + (-33) = -32$ List the pairs whose product is -33 and find the sum.
    $(-1)(33) = -33$ $-1 + 33 = 32$
    $(3)(-11) = -33$ $3 + (-11) = -8$
    $(-3)(11) = -33$ $-3 + 11 = 8$ The correct sum.

    $w^2 - 8w - 33 = (w + 3)(w - 11)$

25. Product: Sum:
$mn = 28$ $m + n = -11$
$(-1)(-28) = 28$ $-1 + (-28) = -29$ List the pairs of negative integers whose product is 28
$(-2)(-14) = 28$ $(-2) + (-14) = -16$ and find the sum.
$(-4)(-7) = 28$ $(-4) + (-7) = -11$
$(-4)(-7) = 28$ $(-4) + (-7) = -11$ The correct sum.

$w^2 - 11w + 28 = (w - 4)(w - 7)$

29. $y^2 + 26y - 27 = 0$
$(y + 27)(y - 1) = 0$ Factor the left side.
$y + 27 = 0$ or $y - 1 = 0$ Zero-product property.
$\quad y = -27$ or $\quad y = 1$

$\{-27, 1\}$

33. $r^2 - 6r - 55 = 0$
$(r - 11)(r + 5) = 0$ Factor the left side.
$r - 11 = 0$ or $r + 5 = 0$ Zero-product property.
$\quad r = 11$ or $\quad r = -5$

$\{-5, 11\}$

37. Product: Sum:
$mn = -40$ $m + n = -6$
$(1)(-40) = -40$ $1 + (-40) = -39$ List the pairs whose product is -40 and find the sum.
$(-1)(40) = -40$ $-1 + 40 = 39$
$(2)(-20) = -40$ $2 + (-20) = -18$
$(-2)(20) = -40$ $-2 + 20 = 18$
$(4)(-10) = -40$ $4 + (-10) = -6$
$(-4)(10) = -40$ $-4 + 10 = 6$
$(5)(-8) = -40$ $5 + (-8) = -3$
$(-5)(8) = -40$ $-5 + 8 = 3$
$(4)(-10) = -40$ $4 + (-10) = -6$ The correct sum.

$x^2 - 6x - 40 = (x + 4)(x - 10)$

41. Product: Sum:
$mn = -40$ $m + n = -39$
$(1)(-40) = -40$ $1 + (-40) = -39$
$(-1)(40) = -40$ $-1 + 40 = 39$
$(2)(-20) = -40$ $2 + (-20) = -18$
$(-2)(20) = -40$ $-2 + 20 = 18$
$(4)(-10) = -40$ $4 + (-10) = -6$
$(-4)(10) = -40$ $-4 + 10 = 6$
$(5)(-8) = -40$ $5 + (-8) = -3$
$(-5)(8) = -40$ $-5 + 8 = 3$
$(1)(-40) = -40$ $1 + (-40) = -39$ The correct sum.

$w^2 - 39w - 40 = (w + 1)(w - 40)$

45. Product: Sum:
$mn = 34$ $m + n = 19$
$(1)(34) = 34$ $1 + 34 = 35$ List the pairs of positive integers whose product is 34
$(2)(17) = 34$ $2 + 17 = 19$ and find the sum.
$(2)(17 = 34$ $2 + 17 = 19$ The correct sum.

$b^2 + 19b + 34 = (b + 2)(b + 17)$

Exercises 4.4

49. $y^2 + 30y + 56 = 0$
$(y + 2)(y + 28) = 0$ Factor the left side.
$y + 2 = 0$ or $y + 28 = 0$ Zero-product property.
$y = -2$ or $y = -28$

$\{-28, -2\}$

53. $r^2 - 2r - 63 = 0$
$(r - 9)(r + 7) = 0$ Factor the left side.
$r - 9 = 0$ or $r + 7 = 0$ Zero-product property.
$r = 9$ or $r = -7$

$\{-7, 9\}$

57. 1st integer: x
2nd integer: $x + 1$

$x(x + 1) = 272$ The product is 272.
$x^2 + x = 272$ Distribute.
$x^2 + x - 272 = 0$ Subtract 272 from both sides.
$(x + 17)(x - 16) = 0$ Factor the left side.
$x + 17 = 0$ or $x - 16 = 0$ Zero-product property.
$x = -17$ or $x = 16$

Since $x > 0$, $x = 16$ and
$x + 1 = 16 + 1 = 17$.
The integers are 16 and 17.

61. $w = 144 - 7t - t^2$ Formula.
$100 = 144 - 7t - t^2$ Replace w with 100.
$t^2 + 7t - 44 = 0$ Add t^2 and $7t$ to both sides. Subtract 144 from both sides.
$(t + 11)(t - 4) = 0$ Factor the left side.
$t + 11 = 0$ or $t - 4 = 0$ Zero-product property.
$t = -11$ or $t = 4$

Since $t > 0$, $t = 4$ sec.

65. One possible answer: Their product is c and their sum is b.

69. Area of   −   Area of   =   Area of
    Entire   Unshaded   Shaded
    Rectangle   Rectangle   Region

$(x + 6)(x + 4) - (4)(6) = A$ Area of Rectangle = (width)(length).
$x^2 + 10x + 24 - 24 = A$ Multiply using FOIL.
$x^2 + 10x = A$ Combine like terms.
$x(x + 10) = A$ Factor.

73. $(2a - 7b)(5a - 4b)$
$= 10a^2 - 8ab - 35ab + 28b^2$ Multiply using FOIL.
$= 10a^2 - 43ab + 28b^2$ Combine like terms.

77. $(22x^2 - 3x + 17) - (12x^2 + 5x - 32) - (x^2 + 25x - 8)$
    $= (22x^2 - 3x + 17) + (-12x^2 - 5x + 32) + (-x^2 - 25x + 8)$ Change to addition.
    $= (22 - 12 - 1)x^2 + (-3 - 5 - 25)x + (17 + 32 + 8)$ Combine like terms.
    $= 9x^2 - 33x + 57$

EXERCISES 4.5

1. $2x^2 + 17x + 26$

    $mn = 2(26) = 52 \qquad m + n = 17$
    $(1)(52) = 52 \qquad\qquad\quad 53$
    $(2)(26) = 52 \qquad\qquad\quad 18$
    $(4)(13) = 52 \qquad\qquad\quad 17$ The correct numbers.

    $2x^2 + 17x + 26 = 2x^2 + 4x + 13x + 26$ Rewrite the trinomial.
    $\qquad\qquad\qquad\;\; = 2x(x + 2) + 13(x + 2)$ Factor by grouping.
    $\qquad\qquad\qquad\;\; = (x + 2)(2x + 13)$

5. $3x^2 - 19x + 28$

    $mn = 3(28) = 84 \qquad m + n = -19$
    $(-1)(-84) = 84 \qquad\qquad\;\; -85$
    $(-2)(-42) = 84 \qquad\qquad\;\; -44$
    $(-3)(-28) = 84 \qquad\qquad\;\; -31$
    $(-4)(-21) = 84 \qquad\qquad\;\; -25$
    $(-6)(-14) = 84 \qquad\qquad\;\; -20$
    $(-7)(-12) = 84 \qquad\qquad\;\; -19$ The correct numbers.

    $3x^2 - 19x + 28 = 3x^2 - 7x - 12x + 28$ Rewrite the trinomial.
    $\qquad\qquad\qquad\;\; = x(3x - 7) - 4(3x - 7)$ Factor by grouping.
    $\qquad\qquad\qquad\;\; = (3x - 7)(x - 4)$

9. $2x^2 + 71x + 35$

    $mn = 2(35) = 70 \qquad m + n = 71$
    $(1)(70) = 70 \qquad\qquad\quad 71$
    $(2)(35) = 70 \qquad\qquad\quad 37$
    $(5)(14) = 70 \qquad\qquad\quad 19$
    $(7)(10) = 70 \qquad\qquad\quad 17$ The correct numbers.

    $2x^2 + 71x + 35 = 2x^2 + x + 70x + 35$ Rewrite the trinomial.
    $\qquad\qquad\qquad\;\; = x(2x + 1) + 35(2x + 1)$ Factor by grouping.
    $\qquad\qquad\qquad\;\; = (2x + 1)(x + 35)$

13. $11x^2 + 80x + 21$

    $mn = 11(21) = 231 \qquad m + n = 80$
    $(1)(231) = 231 \qquad\qquad\quad 232$
    $(3)(77) = 231 \qquad\qquad\quad\;\; 80$
    $(7)(33) = 231 \qquad\qquad\quad\;\; 40$
    $(11)(21) = 231 \qquad\qquad\;\;\; 33$ The correct numbers.

    $11x^2 + 80x + 21 = 11x^2 + 3x + 77x + 21$ Rewrite the trinomial.
    $\qquad\qquad\qquad\quad\; = x(11x + 3) + 7(11x + 3)$ Factor by grouping.
    $\qquad\qquad\qquad\quad\; = (11x + 3)(x + 7)$

Exercises 4.5

17. $20x^2 + 23x - 7$

    $mn = 20(-7) = -140 \quad m + n = 23$
      $(-1)(140)$ 139   Use factors where the positive number has the largest
      $(-2)(70)$ 68   absolute value as the sum, 23 is positive.
      $(-4)(35)$ 31
      $(-5)(28)$ 23   The correct numbers.
      $(-7)(20)$ 14
      $(-10)(14)$ 4

    $20x^2 + 23x - 7 = 20x^2 - 5x + 28x - 7$
                           $= 5x(4x - 1) + 7(4x - 1)$ Rewrite the trinomial.
                           $= (4x - 1)(5x + 7)$ Factor by grouping.

21. $16x^2 + 10x + 1$

    $mn = 16(1) = 16 \quad m + n = 10$
      $(1)(16) = 16$ 17
      $(2)(8) = 16$ 10   The correct numbers.
      $(4)(4) = 16$ 8

    $16x^2 + 10x + 1 = 16x^2 + x + 8x + 1$ Rewrite the trinomial.
                             $= 2x(8x + 1) + 1(8x + 1)$ Factor by grouping.
                             $= (8x + 1)(2x + 1)$

25. $4x^2 - 59x - 15$

    $mn = 4(-15) = -60 \quad m + n = -59$
      $(1)(-60) = -60$ -59   The correct numbers.
      $(2)(-30) = -60$ -28
      $(3)(-20) = -60$ -17   Use factors where the negative number has the largest
      $(4)(-15) = -60$ -11   absolute value as the sum, -59 is negative.
      $(5)(-12) = -60$ -7
      $(6)(-10) = -60$ -4

    $4x^2 - 59x - 15 = 4x^2 + x - 60x - 15$
                          $= x(4x + 1) - 15(4x + 1)$ Rewrite the trinomial.
                          $= (4x + 1)(x - 15)$ Factor by grouping.

29. $15x^2 + 7x - 22$

    $mn = 15(-22) = -330 \quad m + n = 7$
      $(-1)(330) = -330$ 329   Use factors where the positive number has the largest
      $(-2)(165) = -330$ 163   absolute value as the sum, 7 is positive.
      $(-3)(110) = -330$ 107
      $(-5)(66) = -330$ 61
      $(-6)(55) = -330$ 49
      $(-10)(33) = -330$ 23
      $(-11)(30) = -330$ 19
      $(-15)(22) = -330$ 7   The correct numbers.

    $15x^2 + 7x - 22 = 15x^2 - 15x + 22x - 22$ Rewrite the trinomial.
                           $= 15x(x - 1) + 22(x - 1)$ Factor by grouping.
                           $= (x - 1)(15x + 22)$

33. $10x^2 - 13x - 3 = 0$
$(5x + 1)(2x - 3) = 0$ Factor the left side.
$5x + 1 = 0$ or $2x - 3 = 0$ Zero-factor property.
    $5x = -1$ or $2x = 3$
      $x = -\dfrac{1}{5}$ $x = \dfrac{3}{2}$

$\left\{-\dfrac{1}{5}, \dfrac{3}{2}\right\}$

37. $8x^2 - 34x + 21$

    $mn = 8(21) = 168$ $m + n = -34$
    $(-1)(-168) = 168$ -169
    $(-2)(-84) = 168$ -86
    $(-3)(-56) = 168$ -59
    $(-4)(-42) = 168$ -46
    $(-6)(-28) = 168$ -34 The correct numbers.
    $(-7)(-24) = 168$ -31
    $(-8)(-21) = 168$ -29
    $(-12)(-14) = 168$ -26

    $8x^2 - 34x + 21 = 8x^2 - 6x - 28x + 21$ Rewrite the trinomial.
                           $= 2x(4x - 3) - 7(4x - 3)$ Factor by grouping.
                           $= (4x - 3)(2x - 7)$

41. $30x^2 + 425x + 70$
    $= 5(6x^2 + 85x + 14)$ Factor out the common factor 5.

    $mn = 6(14) = 84$ $m + n = 85$ The correct numbers.
    $(1)(84) = 84$ 85
    $(2)(42) = 84$ 44
    $(3)(28) = 84$ 31
    $(4)(21) = 84$ 25
    $(6)(14) = 84$ 20
    $(7)(12) = 84$ 19

    $5(6x^2 + 85x + 14) = 5(6x^2 + x + 84x + 14)$ Rewrite the trinomial.
                                  $= 5[x(6x + 1) + 14(6x + 1)]$ Factor by grouping.
                                  $= 5(6x + 1)(x + 14)$

45. $35x^2 - 39x + 10$

    $mn = 35(10) = 350$ $m + n = -39$
    $(-1)(-350) = 350$ -351
    $(-2)(-175) = 350$ -177
    $(-5)(-70) = 350$ -75
    $(-10)(-35) = 350$ -45
    $(-14)(-25) = 350$ -39 The correct numbers.

    $35x^2 - 39x + 10 = 35x^2 - 14x - 25x + 10$ Rewrite the trinomial.
                                $= 7x(5x - 2) - 5(5x - 2)$ Factor by grouping.
                                $= (5x - 2)(7x - 5)$

Exercises 4.5

49. $18x^2 - 34xy + 15y^2$

$$\begin{array}{ll} mn = 18(15) = 270 & m + n = -37 \\ (-1)(-270) = 270 & -271 \\ (-2)(-135) = 270 & -137 \\ (-3)(-90) = 270 & -93 \\ (-5)(-54) = 270 & -59 \\ (-6)(-45) = 270 & -51 \\ (-9)(-30) = 270 & -39 \\ (-10)(-27) = 270 & -37 \qquad \text{The correct numbers.} \\ (-15)(-18) = 270 & -33 \end{array}$$

$18x^2 - 37xy + 15y^2$
$= 18x^2 - 10xy - 27xy + 15y^2$ \qquad Rewrite the trinomial.
$= 2x(9x - 5y) - 3y(9x - 5y)$ \qquad Factor by grouping.
$= (9x - 5y)(2x - 3y)$

53. $2x^2 - 33x - 35 = 0$
$(2x - 35)(x + 1) = 0$ \qquad Factor the left side.
$2x - 35 = 0 \quad$ or $\quad x + 1 = 0$ \qquad Zero-product property.
$2x = 35 \quad$ or $\quad x = -1$
$x = \dfrac{35}{2}$

$\left\{-1, \dfrac{35}{2}\right\}$

57. $i = 2t^2 - 5t + 25$ \qquad Formula.
$193 = 2t^2 - 5t + 25$ \qquad Replace i with 193.
$0 = 2t^2 - 5t - 168$ \qquad Subtract 193 from both sides.
$0 = (2t - 21)(t + 8)$ \qquad Factor the right side.
$2t - 21 = 0 \quad$ or $\quad t + 8 = 0$ \qquad Zero-product property.
$2t = 21 \quad$ or $\quad t = -8$
$t = \dfrac{21}{2}$

Since $t > 0$, $t = \dfrac{21}{2}$ sec.

61. length: x
width: $2x - 10$ \qquad 10 less than twice the width.

$wl = A$ \qquad Formula.
$(2x - 10)(x) = 72$ \qquad Replace w with $2x - 10$, l with x and A with 72.
$2x^2 - 10x = 72$ \qquad Multiply.
$2x^2 - 10x - 72 = 0$ \qquad Subtract 72 from both sides.
$2(x - 9)(x + 5) = 0$ \qquad Factor the left side.
$x - 9 = 0 \quad$ or $\quad x + 5 = 0$ \qquad Zero-product property.
$x = 9 \quad$ or $\quad x = -5$

Since $x > 0$, $x = 9$ and $2x - 10 = 2(9) - 10 = 8$.
The dimensions are 9 m by 8 m.

65. age today: x
age two years ago: $x - 2$

$$8x^2 - 25(x - 2) = 188 \qquad \text{$8x^2$: eight times square of age today.}$$
$$ \qquad 25(x - 2)\text{: 25 times age two years ago.}$$
$$8x^2 - 25x + 50 = 188 \qquad \text{Multiply.}$$
$$8x^2 - 25x - 138 = 0 \qquad \text{Subtract 188 from both sides.}$$
$$(8x + 23)(x - 6) = 0 \qquad \text{Factor the left side.}$$
$$8x + 23 = 0 \quad \text{or} \quad x - 6 = 0 \qquad \text{Zero-product property.}$$
$$8x = -23 \quad \text{or} \quad x = 6$$
$$x = -\frac{23}{8}$$

Since $x > 0$, $x = 6$ years old.

69. One possible answer: It is helpful to rewrite $ax^2 + bx + c$ as $ax^2 + mx + nx + c$ when $a \neq 1$.

73. $90x^6 + 57x^4 - 12x^2$
$= 3x^2(30x^4 + 19x^2 - 4)$ — Factor out the common factor $3x^2$.

$mn = 30(-4) = -120$	$m + n = 19$
$(-1)(120) = -120$	119
$(-2)(60) = -120$	58
$(-3)(40) = -120$	37
$(-4)(30) = -120$	26
$(-5)(24) = -120$	19 — The correct numbers.
$(-6)(20) = -120$	14
$(-8)(15) = -120$	7
$(-10)(12) = -120$	2

$ 3x^2(30x^4 + 19x^2 - 4)$
$= 3x^2(30x^4 - 5x^2 + 24x^2 - 4) \qquad$ Rewrite the trinomial.
$= 3x^2[5x^2(6x^2 - 1) + 4(6x^2 - 1)] \qquad$ Factor by grouping.
$= 3x^2(6x^2 - 1)(5x^2 + 4)$

77. $(2x - 5)(2x + 5) = (2x)^2 - 5^2 \qquad$ Multiplication of conjugate results in the
$ = 4x^2 - 25 \qquad$ difference of squares.

81. $(5x - 1)^2 = (5x)^2 - 2(5x)(1) + 1^2 \qquad$ Square the first term, subtract twice the
$ = 25x^2 - 10x + 1 \qquad$ product of the two terms and add the
$ \qquad$ square of the second term.

85. $(x - 5)(x + 14) - (x - 12)(x + 24) = 16$
$x^2 + 9x - 70 - (x^2 + 12x - 288) = 16 \qquad$ Multiply using FOIL.
$x^2 + 9x - 70 + (-x^2 - 12x + 288) = 16 \qquad$ Change to addition.
$-3x + 218 = 16 \qquad$ Combine like terms.
$-3x = -202 \qquad$ Subtract 218 from both sides.
$x = \dfrac{202}{3} \qquad$ Divide both sides by -3.

$\left\{\dfrac{202}{3}\right\}$

EXERCISES 4.6

1. $x^2 - 9 = (x)^2 - (3)^2$

    $= (x + 3)(x - 3)$

    The first term is the square of x and the second term is the square of 3.
    The factors are the sum and difference of the terms that are squared.

5. $y^3 - 125$
    $= y^3 - 5^3$
    $= (y - 5)[(y)^2 + (y)(5) + (5)^2]$
    $= (y - 5)(y^2 + 5y + 25)$

    Identify the terms that are cubed.
    Substitute into the formula $a^3 - b^3$
    $= (a - b)(a^2 + ab + b^2)$.
    Simplify.

9. $w^2 + 4w + 4$
    $= (w)^2 + 2(w)(2) + (2)^2$
    $= (w + 2)^2$

    Check to see that twice the product of w and 2 is $4w$.
    Write the square of the sum of the two terms.

13. $x^2 - 64 = 0$
    $(x + 8)(x - 8) = 0$
    $x + 8 = 0$ or $x - 8 = 0$
    $x = -8$ or $x = 8$

    $\{-8, 8\}$

    Factor the left side.
    Zero-product property.

17. $w^2 + 16w + 64 = 0$
    $(w + 8)^2 = 0$
    $w + 8 = 0$ or $w + 8 = 0$
    $w = -8$ or $w = -8$

    $\{-8\}$

    Factor the left side.
    Zero-product property.
    A double root.

21. $9x^2 - 4 = (3x)^2 - (2)^2$

    $= (3x + 2)(3x - 2)$

    The first term is the square of $3x$ and the second term is the square of 2.
    The factors are the sum and difference of the terms that are squared.

25. $8x^3 - 1$
    $= (2x)^3 - (1)^3$
    $= (2x - 1)[(2x)^2 + (2x)(1) + (1)^2]$
    $= (2x - 1)(4x^2 + 2x + 1)$

    Identify the terms that are cubed.
    Substitute into the formula $a^3 - b^3$
    $= (a - b)(a^2 + ab + b^2)$.
    Simplify.

29. $4w^2 + 28w + 49$
    $= (2w)^2 + 2(2w)(7) + (7)^2$
    $= (2w + 7)^2$

    Check to see that twice the product of $2w$ and 7 is $28w$.
    Write the square of the sum of the two terms.

33. $25x^2 - 36 = 0$
    $(5x + 6)(5x - 6) = 0$
    $5x + 6 = 0$ or $5x - 6 = 0$
    $5x = -6$ or $5x = 6$
    $x = -\dfrac{6}{5}$ or $x = \dfrac{6}{5}$

    $\left\{-\dfrac{6}{5}, \dfrac{6}{5}\right\}$

    Factor the left side.
    Zero-product property.

37. $4w^2 + 20w + 25 = 0$
    $(2w + 5)^2 = 0$ Factor the left side.
    $2w + 5 = 0$ or $2w + 5 = 0$ Zero-product property.
        $2w = -5$ or $2w = -5$
        $w = -\dfrac{5}{2}$ or $w = -\dfrac{5}{2}$ A double root.
    $\left\{-\dfrac{5}{2}\right\}$

41. Prime   The sum of two squares cannot be factored if there is no common monomial factor.

45. $64x^3 - 27$
    $= (4x)^3 - (3)^3$ Identify the terms that are cubed.
    $= (4x - 3)[(4x)^2 + (4x)(3) + (3)^2]$ Substitute into the formula $a^3 - b^3 = (a - b)(a^2 + ab + b^2)$.
    $= (4x - 3)(16x^2 + 12x + 9)$ Simplify.

49. $169x^2 - 286x + 121$
    $= (13x)^2 + 2(13x)(-11) + (-11)^2$ Since the middle term is negative we write 121 as $(-11)^2$ instead of 11^2.
    $= [13x + (-11)]^2$ Write the square of the sum of the two terms.
    $= (13x - 11)^2$ Simplify.

53. $11x^2 + 3 = 2x^2 + 19$
    $9x^2 - 16 = 0$ Subtract $2x^2$ and 19 from both sides.
    $(3x + 4)(3x - 4) = 0$ Factor the left side.
    $3x + 4 = 0$ or $3x - 4 = 0$ Zero-product property.
        $3x = -4$ or $3x = 4$
        $x = -\dfrac{4}{3}$ or $x = \dfrac{4}{3}$
    $\left\{-\dfrac{4}{3}, \dfrac{4}{3}\right\}$

57. $y^2 - 5y + 10 = 26 - 5y$
    $y^2 - 16 = 0$ Subtract 26 from both sides and add $5y$ to both sides.
    $(y + 4)(y - 4) = 0$ Factor the left side.
    $y + 4 = 0$ or $y - 4 = 0$ Zero-product property.
        $y = -4$ or $y = 4$
    $\{-4, 4\}$

61. $z^2 - z - 1 = 77z - 8z^2 - 170$
    $9z^2 - 78z + 169 = 0$ Add $8z^2$ and 170 to both sides. Subtract $77z$ from both sides.
    $(3z - 13)^2 = 0$ Factor the left side.

Exercises 4.6

$$3z - 13 = 0 \quad \text{or} \quad 3z - 13 = 0$$
$$3z = 13 \quad \text{or} \quad 3z = 13$$
$$z = \frac{13}{3} \quad \text{or} \quad z = \frac{13}{3}$$

Zero-product property.

A double root.

$\left\{\frac{13}{3}\right\}$

65. age today: x
    age three years ago: $x - 3$

$$4x^2 - 10(x - 3) = 5$$
$$4x^2 - 10x + 30 = 5$$
$$4x^2 - 10x + 25 = 0$$
$$(2x - 5)^2 = 0$$
$$2x - 5 = 0$$
$$2x = 5$$
$$x = \frac{5}{2}$$

$4x^2$: four times the square of today's age.
$10(x - 3)$: ten times age three years ago.
Multiply.
Subtract 5 from both sides.
Factor the left side.
Zero-product property.

Double root.

Krista's age: $\frac{5}{2} = 2\frac{1}{2}$ yr

69. $100 - 9 = (10)^2 (3)^2$
    $= (10 + 3)(10 - 3)$
    $= (13)(7)$

Write 100 as 10^2 and 9 as 3^2.
The factors are the sum and difference of the terms that are squared.
Simplify.

7 and 13

73. One possible answer: In a perfect square trinomial the first and last terms are both perfect squares and the middle term is twice the product of the terms that are squared in the first and last terms.

77. $(x - y)^2 - 49 = (x - y)^2 - (7)^2$
    $= [(x - y) + 7][(x - y) - 7]$
    $= (x - y + 7)(x - y - 7)$

The first term is the square of $x - y$ and the second term is the square of 7.
The factors are the sum and difference of the terms that are squared.
Simplify.

81. $(x - 4)(x + 9) = x^2 + 9x - 4x - 36$
    $= x^2 + 5x - 36$

Multiply using FOIL.
Combine like terms.

85. $(x - 2y)(x + 2y) = (x)^2 - (2y)^2$
    $= x^2 - 4y^2$

When multiplying conjugates the result is the difference of squares.
Simplify.

89. $(2y - 5)(7y + 8) = 14y^2 + 16y - 35y - 40$
    $= 14y^2 - 19y - 40$

Multiply using FOIL.
Combine like terms.

EXERCISES 4.7

1. $6a^2 + 8ab = 2a(3a + 4b)$ The GCF is $2a$.

5. $16a^2b^4 - 25 = (4ab^2) - (5)^2$
   $= (4ab^2 + 5)(4ab^2 - 5)$

   $16a^2b^4w$ is the square of $4ab^2$ and 25 is the square of 5. The factors are the sum and difference of the terms that are squared.

9. $5c^2 - 5cd + 3cd - 3d^2$
   $= (5c^2 - 5cd) + (3cd - 3d^2)$
   $= 5c(c - d) + 3d(c - d)$
   $= (c - d)(5c + 3d)$

   Group the first two terms and the last two terms.
   Factor $5c$ out of the first group and $3d$ out of the second group.
   Factor out $c - d$.

13. $7x^3 - 7$
    $= 7(x^3 - 1)$
    $= 7[(x)^3 - (1)^3]$

    $= 7\{(x - 1)[(x)^2 + (x)(1) + (1)^2]\}$
    $= 7(x - 1)(x^2 + x + 1)$

    The GCF is 7.
    Identify the terms that are cubed.
    Substitute into the formula $a^3 - b^3$
    $= (a - b)(a^2 + ab + b^2)$.
    Simplify.

17. $2r^2 - 28r + 98 = 0$
    $2(r^2 - 14r + 49) = 0$
    $2(r - 7)^2 = 0$
    $r - 7 = 0$ or $r - 7 = 0$
    $r = 7$ or $r = 7$
    $\{7\}$

    The GCF is 2.
    Factor the left side.
    Zero-product property.
    A double root.

21. $8x^4y - 16x^2y^3 - 4x^2y$
    $= 4x^2y(2x^2 - 4y^2 - 1)$ The GCF is $4x^2y$.

25. $24m^2 + 8mt - 9mt - 3t^2$
    $= (24m^2 + 8mt) + (-9mt - 3t^2)$
    $= 8m(3m + t) - 3t(3m + t)$
    $= (3m + t)(8m - 3t)$

    Group the first two terms and the last two terms.
    Factor $8m$ out of the first group and $-3t$ out of the second group.
    Factor out $3m + t$.

29. $2aw^2 + 44aw + 242a$
    $= 2a(w^2 + 22w + 121)$
    $= 2a[(w)^2 + 2(w)(11) + (11)^2]$
    $= 2a(w + 11)^2$

    The GCF is $2a$.
    Check to see that twice the product of w and 11 is $22w$.
    Write the square of the sum of the two terms.

33. $r^2 + 9 = 6r$
    $r^2 - 6r + 9 = 0$
    $(r - 3)^2 = 0$
    $r - 3 = 0$ or $r - 3 = 0$
    $r = 3$ or $r = 3$
    $\{3\}$

    Subtract $6r$ from both sides.
    Factor the left side.
    Zero-product property.
    A double root.

37. $3t^2 = -12 - 20t$
    $3t^2 + 20t + 12 = 0$
    $(3t + 2)(t + 6) = 0$

    Add 12 and $20t$ to both sides.
    Factor the left side.

Exercises 4.7

$3t + 2 = 0$ or $t + 6 = 0$ Zero-product property.
$3t = -2$ or $t = -6$

$t = -\dfrac{2}{3}$

$\left\{-6, -\dfrac{2}{3}\right\}$

41. $5x^3 - 5xy^2 + 2x^2y - 2y^3$
$= (5x^3 - 5xy^2) + (2x^2y - 2y^3)$ Group the first two terms and the last two terms.
$= 5x(x^2 - y^2) + 2y(x^2 - y^2)$ Factor $5x$ out of the first group and $2y$ out of the second group.
$= (x^2 - y^2)(5x + 2y)$ Factor out $x^2 - y^2$.
$= (x + y)(x - y)(5x + 2y)$ Factor the difference of squares.

45. $8x^3 + 72x^2 + 112x$
$= 8x(x^2 + 9x + 14)$ The GCF is $8x$.
$= 8x(x + 7)(x + 2)$ Product: $mn = 14$.
    Sum: $m + n = 9$.
    The correct numbers are 7 and 2.

49. $72a^3b^3 - 66a^2b^2c - 30abc^2$
$= 6ab(12a^2b^2 - 11abc - 5c^2)$
$= 6ab(12a^2b^2 - 15abc + 4abc - 5c^2)$ The GCF is $6ab$.
    $mn = 12(-5) = -60$.
    $m + n = -11$
    The correct numbers are 4 and -15.
    Rewrite the binomial.

$= 6ab[3ab(4ab - 5c) + c(4ab - 5c)]$
$= 6ab(4ab - 5c)(3ab + c)$ Factor by grouping.

53. $4a^3b^2c - 32b^2c^4$
$= 4b^2c(a^3 - 8c^3)$ The GCF is $4b^2c$.
$= 4b^2c[(a)^3 - (2c)^3]$ Identify the terms that are cubed.
$= 4b^2c\{(a - 2c)[(a)^2 + (a)(2c) + (2c)^2]\}$ Substitute into the formula $a^3 - b^3 = (a - b)(a^2 + ab + b^2)$.
$= 4b^2c(a - 2c)(a^2 + 2ac + 4c^2)$ Simplify.

57. $(x - 4)^2 = 25$
$x^2 - 8x + 16 = 25$ Multiply using FOIL.
$x^2 - 8x - 9 = 0$ Subtract 25 from both sides.
$(x - 9)(x + 1) = 0$ Factor the left side.
$x - 9 = 0$ or $x + 1 = 0$ Zero-product property.
$x = 9$ or $x = -1$

$\{-1, 9\}$

61. $(y - a)(y - a) + (y + a)(y + a) = 10a^2$
$y^2 - 2ay + a^2 + y^2 + 2ay + a^2 = 10a^2$ Multiply using FOIL.
$2y^2 + 2a^2 = 10a^2$ Combine like terms.
$2y^2 - 8a^2 = 0$ Subtract $10a^2$ from both sides.
$2(y^2 - 4a^2) = 0$ The GCF is 2.
$2(y + 2a)(y - 2a) = 0$ Factor the left side.
$y + 2a = 0$ or $y - 2a = 0$ Zero-product property.
$y = -2a$ or $y = 2a$

65. frame's width: x

$A = wl$
$99 = (5 + 2x)(7 + 2x)$ Formula.
　　　　　　　　　　　　　　　The width of photo and frame is $5 + x + x$.
　　　　　　　　　　　　　　　The length of photo and frame is $7 + x + x$.
$99 = 35 + 24x + 4x^2$ Multiply using FOIL.
$0 = 4x^2 + 24x - 64$ Subtract 99 from both sides.
$0 = 4(x^2 + 6x - 16)$ The GCF is 4.
$0 = 4(x + 8)(x - 2)$ Factor the left side.
$x + 8 = 0$　or　$x - 2 = 0$　Zero-product property.
　$x = -8$　or　　$x = 2$

Since $x > 0$, $x = 2$ inches.

69. width: x $2w + 2l = 400$
　　length: $200 - x$ $2l = 400 - 2w$
　　　　　　　　　　　　　　　　　$l = 200 - w.$
　　　　　　　　　　　　　　　Formula.
$A = wl$ Replace A with 9900, w with x and l with $200 - x$.
$9900 = x(200 - x)$ Multiply.
$9900 = 200x - x^2$ Add x^2 to both sides.
$x^2 - 200x + 9900 = 0$ Subtract $200x$ from both sides.
　　　　　　　　　　　　　　　Factor the left side.
$(x - 110)(x - 90) = 0$ Zero-product property.
$x - 110 = 0$　　or　$x - 90 = 0$
　$x = 110$　　or　　$x = 90$
$200 - x = 200 - 110$　$200 - x = 200 - 90$
　　　　　$= 90$　　　　　　　　$= 110$

The dimensions are 110 ft by 90 ft.

73. One possible answer: First look for a common factor and if there is one, use the distributive property to factor it out. If any of the factors are binomials check to see if they are the difference of two squares or sum or difference of two cubes. If any of the factors are trinomials, try to factor by appropriate method. If any factors contain 4 or more terms, factor by grouping. In the end make make sure that none of the factors can be factored further.

77.　$x^4 - 13x^2 + 36$ $mn = 36;\ m + n = -13.$
　　$= (x^2 - 9)(x^2 - 4)$ Write 9 as 3^2 and 4 as 2^2.
　　$= [(x)^2 - (3)^2][(x)^2 - (2)^2]$ Factor the difference of two squares.
　　$= (x + 3)(x - 3)(x + 2)(x - 2)$

81.　$(a - b + 3)(2a + 3b - 5)$
　　$= a(2a + 3b - 5) - b(2a + 3b - 5)$ Distributive property.
　　　$+ 3(2a + 3b - 5)$
　　$= 2a^2 + 3ab - 5a - 2ab - 3b^2 + 5b$ Multiply.
　　　$+ 6a + 9b - 15$
　　$= 2a^2 + ab + a - 15 + 14b - 3b^2$ Combine like terms.

85.　　　$(3x - 1)(x^2 - 5x - 8) = 3(x^3 - 2) - 16x^2$
　$3x^3 - 15x^2 - 24x - x^2 + 5x + 8 = 3x^3 - 6 - 16x^2$ Multiply.
　　　　$3x^3 - 16x^2 - 19x + 8 = 3x^3 - 6 - 16x^2$ Combine like terms.

$$-19x + 8 = -6$$
$$-19x = -14$$

Subtract $3x^3$ from both sides. Add $16x^2$ to both sides. Subtract 8 from both sides.

$$x = \frac{14}{19}$$

Divide both sides by -19.

$$\left\{\frac{14}{19}\right\}$$

89.
$$\begin{array}{r} x^2 - 4x + 6 \\ x + 7 \overline{\smash{\big)}\, x^3 + 3x^2 - 22x + 42} \\ \underline{x^3 + 7x^2 } \\ -4x^2 - 22x + 42 \\ \underline{-4x^2 - 28x } \\ 6x + 42 \\ \underline{6x + 42} \end{array}$$

$x^3 \div x = x^2$.
Multiply $x^2(x + 7)$.
Subtract.
$-4x^2 \div x = -4x$.

$6x \div x = 6$.

CHAPTER 4 CONCEPT REVIEW

1. False;   A quadratic equation is a second-degree equation.

2. False;   Either $A = 0$ or $B = 0$.

3. False;   A quadratic equation may have one root (double root).

4. True

5. False;   It is a sum, not a product.

6. True

7. True

8. False;   It is a subtraction problem, not a product.

9. True

10. False;   It is true if $a + b = c$ and $ab = d$.

11. True

12. True

13. False; $x^3 + y^3 = (x + y)(x^2 - xy + y^2)$.

14. False;   The middle term needs to be $2(4x)(5) = 40x$.

15. False; $(x^3 - 1)(x^3 + 1) = x^6 - 1$.

16. False; $ad + bc = f$.

17. True

18. False; $\qquad$ $x^3 - y^3 = (x - y)(x^2 + xy + y^2)$.

19. True

20. False; $\qquad$ $4x^2 + 16y^2 = 4(x^2 + 4y^2)$.

CHAPTER 4 TEST

1. $75a^3 - 270a^2b + 243ab^2$
   $= 3a(25a^2 - 90ab + 81b^2)$ The GCF is $3a$.
   $= 3a[(5a)^2 + 2(5a)(-9b) + (-9b)^2]$ Write $81b^2$ as $(-9b)^2$ since the middle term is negative.
   $= 3a[5a + (-9b)]^2$ Square the sum of the two terms.
   $= 3a(5a - 9b)^2$ Simplify.

2. $ax - bx + 2a - 2b$
   $= (ax - bx) + (2a - 2b)$ Group the first two terms and the last two terms.
   $= x(a - b) + 2(a - b)$ Factor x out of the first group and 2 out of the second group.
   $= (a - b)(x + 2)$ Factor out $a - b$.

3. $x^2 + 20x + 36 = 0$
   $(x + 18)(x + 2) = 0$ Factor the left side.
   $x + 18 = 0$ or $x + 2 = 0$ Zero-product property.
   $x = -18$ or $x = -2$

   $\{-18, -2\}$

4. $x(x - 2) = 0$
   $x = 0$ or $x - 2 = 0$ Zero-product property.
   $x = 2$

   $\{0, 2\}$

5. $x^2 - 24x + 144 = 0$
   $(x - 12)^2 = 0$ Factor the left side.
   $x - 12 = 0$ or $x - 12 = 0$ Zero-product property.
   $x = 12$ or $x = 12$ A double root.

   $\{12\}$

6. $22ab - 44abc$
   $= 22ab(1 - 2c)$ The GCF is $22ab$.

7. $x^2 + 5x = 0$
   $x(x + 5) = 0$ Factor the left side.
   $x = 0$ or $x + 5 = 0$ Zero-product property.
   $x = -5$

   $\{-5, 0\}$

8. $r^2 - 19r + 18$
   $= (r - 18)(r - 1)$ $mn = 18; \;\; m + n = -19$.
   $(-18)(-1) = 18; \;\; -18 + (-1) = -19$.

9. $6x^2 - 5x - 6$
   $= 6x^2 - 9x + 4x - 6$
   $= 3x(2x - 3) + 2(2x - 3)$
   $= (2x - 3)(3x + 2)$

$mn = 6(-6) = -36;\ m + n = -5.\ (-9)(4) = -36;\ -9 + 4 = -5.$
Rewrite the trinomial.
Factor by grouping.

10. $14x^2y + 28xy - 77wy$
    $= 7y(2x^2 + 4x - 11w)$

The GCF is $7y$.

11. $y^2 + 5y - 66$
    $= (y + 11)(y - 6)$

$mn = -66;\ m + n = 5.$
$(11)(-6) = -66;\ 11 + (-6) = 5.$

12. $8x^2 - 4x + 2xy - y$
    $= (8x^2 - 4x) + (2xy - y)$
    $= 4x(2x - 1) + y(2x - 1)$
    $= (2x - 1)(4x + y)$

Group the first two terms and the last two terms.
Factor $4x$ out of the first group and y out of the second group.
Factor out $2x - 1$.

13. $3x^3 - 375y^3$
    $= 3(x^3 - 125y^3)$
    $= 3[(x)^3 - (5y)^3]$
    $= 3(x - 5y)[(x)^2 + (x)(5y) + (5y)^2]$
    $= 3(x - 5y)(x + 5xy + 25y^2)$

The GCF is 3.
Identify the terms that are cubed.
Substitute into the formula $a^3 - b^3 = (a - b)(a^2 + ab + b^2)$.
Simplify.

14. $36b^2c^2 - 1$
    $= (6bc)^2 - (1)^2$
    $= (6bc + 1)(6bc - 1)$

Write $36b^2c^2$ as the square of $6bc$ amd 1 as the square of 1.
The factors are the sum and difference of the terms that are squared.

15. $6t^2 + 7t - 20$

    $= 6t^2 - 8t + 15t - 20$
    $= 2t(3t - 4) + 5(3t - 4)$
    $= (3t - 4)(2t + 5)$

$mn = 6(-20) = -120;\ m + n = 7.$
$(-8)(15) = -120;\ -8 + 15 = 7.$
Rewrite the trinomial.
Factor by grouping.

16. $9y^2 - 100 = 0$
    $(3y + 10)(3y - 10) = 0$
$3y + 10 = 0$ or $3y - 10 = 0$
    $3y = -10$ or $3y = 10$
    $y = -\dfrac{10}{3}$ $y = \dfrac{10}{3}$

$\left\{-\dfrac{10}{3}, \dfrac{10}{3}\right\}$

Factor the left side.
Zero-product property.

17. $6x^2 - 37x + 6 = 0$
   $(6x - 1)(x - 6) = 0$
$6x - 1 = 0$ or $x - 6 = 0$
   $6x = 1$ or $x = 6$

   $x = \dfrac{1}{6}$

$\left\{\dfrac{1}{6}, 6\right\}$

Factor the left side.
Zero-product property.

18. $36x^2 + 84xy + 49y^2$
    $= (6x)^2 + 2(6x)(7y) + (7y)^2$ Check to see that twice the product of $6x$ and $7y$ is $84xy$.
    $= (6x + 7y)^2$ Write the square of the sum of the two terms.

19. 1st number: x
    2nd number: $9 - x$ $x + 9 - x = 9.$

    $x(9 - x) = -90$ The product is -90.
    $9x - x^2 = -90$ Multiply.
    $0 = x^2 - 9x - 90$ Add x^2 to both sides.
        Subtract $9x$ from both sides.
    $0 = (x - 15)(x + 6)$ Factor the right side.
    $x - 15 = 0$ or $x + 6 = 0$ Zero-product property.
    $x = 15$ or $x = -6$
    $9 - x = 9 - 15$ $9 - x = 9 - (-6)$
    $= -6$ $= 15$

    The numbers are -6 and 15.

20. $i = 2t^2 - 5t + 16$ Formula.
    $58 = 2t^2 - 5t + 16$ Replace i with 58.
    $0 = 2t^2 - 5t - 42$ Subtract 58 from both sides.
    $0 = (2t + 7)(t - 6)$ Factor the right side.
    $2t + 7 = 0$ or $t - 6 = 0$ Zero-product property.
    $2t = -7$ or $t = 6$
    $t = -\dfrac{7}{2}$

    Since $t > 0$, $t = 6$ sec.

EXERCISES 5.1

1. $\dfrac{xy}{y^2} = \dfrac{y \cdot x}{y \cdot y}$ Factor the numerator and the denominator.

    $= \dfrac{x}{y}$ Eliminate the common factor y.

5. $\dfrac{30a^3b^4}{5a^3b^2} = \dfrac{5a^3b^2 \cdot 6b^2}{5a^3b^2 \cdot 1}$ Factor the numerator and the denominator.

    $= \dfrac{6b^2}{1}$ Eliminate the common factor $5a^3b^2$.

    $= 6b^2$ Reduce.

9. $\dfrac{3x - 12}{5x - 20} = \dfrac{3(x - 4)}{5(x - 4)}$ Factor the numerator and the denominator.

    $= \dfrac{3}{5}$ Eliminate the common factor $x - 4$.

13. $\dfrac{4m}{r} \cdot \dfrac{5m}{7}$

    $= \dfrac{4m(5m)}{r(7)}$ Since there are no common factors, multiply the numerators and the denominators.

    $= \dfrac{20m^2}{7r}$

17. $\dfrac{1}{2} \cdot \dfrac{-a}{b} \cdot \dfrac{c}{d}$

    $= \dfrac{(1)(-a)(c)}{(2)(b)(d)}$ Since there are no common factors, multiply the numerators and the denominators.

    $= -\dfrac{ac}{2bd}$

21. $\dfrac{18a^2b}{24a^2b^2} = \dfrac{6a^2b \cdot 3}{6a^2b \cdot 4b}$ Factor the numerator and the denominator.

    $= \dfrac{3}{4b}$ Eliminate the common factor $6a^2b$.

25. $\dfrac{2cm + 2m}{3c + 3} = \dfrac{2m(c + 1)}{3(c + 1)}$ Factor the numerator and the denominator.

    $= \dfrac{2m}{3}$ Eliminate the common factor $c + 1$.

29. $\dfrac{6x^2 + 5x - 4}{2x^2 + 11x - 6} = \dfrac{(3x + 4)(2x - 1)}{(x + 6)(2x - 1)}$ Factor the numerator and the denominator.

$\qquad\qquad\qquad = \dfrac{3x + 4}{x + 6}$ Eliminate the common factor $2x - 1$.

33. $\dfrac{a^3 b^c}{x^2 y^3} \cdot \dfrac{x^5 y^4}{a^2 b}$

$= \dfrac{a^2 b \cdot abc}{x^2 y^3 \cdot 1} \cdot \dfrac{x^2 y^3 \cdot x^3 y}{a^2 b \cdot 1}$ Factor.

$= \dfrac{abc}{1} \cdot \dfrac{x^3 y}{1}$ Reduce.

$= \dfrac{abcx^3 y}{1}$ Multiply.

$= abcx^3 y$ Reduce.

37. $\dfrac{2x + 5}{12} \cdot \dfrac{4}{4x + 10}$

$= \dfrac{2x + 5}{4 \cdot 3} \cdot \dfrac{4}{2(2x + 5)}$ Factor.

$= \dfrac{1}{3} \cdot \dfrac{1}{2}$ Reduce.

$= \dfrac{1}{6}$ Multiply.

41. $\dfrac{2x + 3}{x - 1} \cdot \dfrac{3x - 2}{x + 4}$

$= \dfrac{(2x + 3)(3x - 2)}{(x - 1)(x + 4)}$ There are no common factors, so multiply.

$= \dfrac{6x^2 + 5x - 6}{x^2 + 3x - 4}$ FOIL the numerator and the denominator.

45. $\dfrac{a^2 + 7a + 12}{a^2 + 5a + 6} \cdot \dfrac{a^2 + 8a + 15}{a^2 + 5a + 4}$

$= \dfrac{(a + 3)(a + 4)}{(a + 2)(a + 3)} \cdot \dfrac{(a + 5)(a + 3)}{(a + 4)(a + 1)}$ Factor.

$= \dfrac{1}{a + 2} \cdot \dfrac{(a + 5)(a + 3)}{a + 1}$ Reduce.

$= \dfrac{(a + 5)(a + 3)}{(a + 2)(a + 1)}$ Multiply.

$= \dfrac{a^2 + 8a + 15}{(a + 2)(a + 1)}$ FOIL the numerator.

Exercises 5.1

49. $\dfrac{2a^2 - 3a - 2}{a^2 - 2a - 15} \cdot \dfrac{3a^2 - 20a - 7}{3a^2 - 5a - 2}$

   $= \dfrac{(2a + 1)(a - 2)}{(a - 5)(a + 3)} \cdot \dfrac{(3a + 1)(a - 7)}{(3a + 1)(a - 2)}$ Factor.

   $= \dfrac{2a + 1}{(a - 5)(a + 3)} \cdot \dfrac{a - 7}{1}$ Reduce.

   $= \dfrac{(2a + 1)(a - 7)}{(a - 5)(a + 3)}$ Multiply.

53. $\dfrac{x^2 + 5x - 14}{x^2 - x - 2} \cdot \dfrac{3x^2 + x - 2}{2x^2 + 13x - 7}$

   $= \dfrac{(x + 7)(x - 2)}{(x + 1)(x - 2)} \cdot \dfrac{(3x - 2)(x + 1)}{(2x - 1)(x + 7)}$ Factor.

   $= \dfrac{1}{1} \cdot \dfrac{3x - 2}{2x - 1}$ Reduce.

   $= \dfrac{3x - 2}{2x - 1}$ Multiply.

57. $\dfrac{ab - 3b^2}{2a^2 - 5ab} \cdot \dfrac{a^2 + 2ab + b^2}{a^2b - 2ab^2 - 3b^3} \cdot \dfrac{2a^2 - 5ab}{a^2 - b^2}$

   $= \dfrac{b(a - 3b)}{a(2a - 5b)} \cdot \dfrac{(a + b)^2}{b(a - 3b)(a + b)} \cdot \dfrac{a(2a - 5b)}{(a + b)(a - b)}$ Factor.

   $= \dfrac{1}{1} \cdot \dfrac{1}{1} \cdot \dfrac{1}{a - b}$ Reduce.

   $= \dfrac{1}{a - b}$ Multiply.

61. $x^2 - 3x + 2 = 0$ Find the values of x that cause the denominator to equal 0.
      $(x - 2)(x - 1) = 0$ Factor.
      $x - 2 = 0$ or $x - 1 = 0$ Zero-product property.
        $x = 2$ or $x = 1$

   Linda wins since $x = 2$ and $x = 1$ cause the expression to not equal a real number.

65. $P = \dfrac{1}{t} \cdot \dfrac{1}{t} \cdot \dfrac{1}{t} \cdot \dfrac{1}{t} \cdot \dfrac{1}{t} \cdot \dfrac{1}{t} \cdot \dfrac{1}{t}$

   $= \dfrac{1 \cdot 1 \cdot 1 \cdot 1 \cdot 1 \cdot 1 \cdot 1}{t \cdot t \cdot t \cdot t \cdot t \cdot t \cdot t}$ Multiply.

   $= \dfrac{1}{t^7}$ Add the exponents.

$$P = \frac{1}{t^7}$$ Formula.

$$P = \frac{1}{5^7} = \frac{1}{78125}$$ Replace t with 5.
$5^7 = 78125$.

69. $c = \dfrac{wrt}{1000}$ Formula.

$c = \dfrac{(2000)(0.12)(4.5)}{1000}$ Replace w with 2000. r with 0.12 and t with 4.5.

$c = \dfrac{(2)(0.12)(4.5)}{1}$ Reduce.

$c = 1.08$ Multiply.

$1.08

73. $\dfrac{8y^3 - 1}{7 - y} \cdot \dfrac{y^2 - 49}{4y^2 - 1}$

$= \dfrac{(2y - 1)(4y^2 + 2y + 1)}{-(y - 7)} \cdot \dfrac{(y + 7)(y - 7)}{(2y + 1)(2y - 1)}$ Factor.

$= \dfrac{4y^2 + 2y + 1}{-1} \cdot \dfrac{y + 7}{2y + 1}$ Reduce.

$= -\dfrac{(y + 7)(4y^2 + 2y + 1)}{2y + 1}$ Multiply.

Restrictions:
$y - 7 \neq 0;\quad 2y + 1 \neq 0;\quad 2y - 1 \neq 0$
$y \neq 7;\quad 2y \neq -1;\quad 2y \neq 1$
$\qquad\qquad y \neq -\dfrac{1}{2};\quad y \neq \dfrac{1}{2}$

81. $(-5)^4 = (-5)(-5)(-5)(-5)$ -5 is used as a factor 4 times.
$= 625$ Multiply.

85. $(x - 12)(x + 16) = 0$
$x - 12 = 0$ or $x + 16 = 0$ Zero-product property.
$x = 12$ or $x = -16$

$\{-16, 12\}$

EXERCISES 5.2

1. $\dfrac{3c}{y} \div \dfrac{4x}{d} = \dfrac{3c}{y} \cdot \dfrac{d}{4x}$ Multiply by the reciprocal of the divisor.

$= \dfrac{3cd}{4xy}$ Multiply.

Exercises 5.2

5. $\dfrac{4yz}{-3} \div \dfrac{-w}{8y} = \dfrac{4yz}{-3} \cdot \dfrac{8y}{-w}$ Multiply by the reciprocal of the divisor.

$= \dfrac{32y^2z}{3w}$ Multiply.

9. $\dfrac{c}{d^2} \div \dfrac{c^2}{f} = \dfrac{c}{d^2} \cdot \dfrac{f}{c^2}$ Multiply by the reciprocal of the divisor.

$= \dfrac{1 \cdot f}{d^2 \cdot c}$ Reduce and multiply.

$= \dfrac{f}{cd^2}$

13. $\dfrac{m+n}{5} \div \dfrac{m+n}{7} = \dfrac{m+n}{5} \cdot \dfrac{7}{m+n}$ Multiply by the reciprocal of the divisor.

$= \dfrac{1 \cdot 7}{5 \cdot 1}$ Reduce and multiply.

$= \dfrac{7}{5}$

17. $\dfrac{5a+3b}{x} \div \dfrac{3a+c}{9ab} = \dfrac{5a+3b}{x} \cdot \dfrac{9ab}{3a+c}$ Multiply by the reciprocal of the divisor.

$= \dfrac{45a^2b + 27ab^2}{3ax + cx}$ Multiply.

21. $\dfrac{12x^3y}{8xy^7} \div \dfrac{7x^5y^7}{6x^2y} = \dfrac{12x^3y}{8xy^7} \cdot \dfrac{6x^2y}{7x^5y^7}$ Multiply by the reciprocal of the divisor.

$= \dfrac{3 \cdot 3}{7 \cdot x \cdot y^5 \cdot y^7}$ Reduce and multiply.

$= \dfrac{9}{7xy^{12}}$

25. $\dfrac{10a^2 - 5a}{11a^3 + 22a^2} \div \dfrac{2a-1}{a^2+2a}$

$= \dfrac{10a^2 - 5a}{11a^3 + 22a^2} \cdot \dfrac{a^2+2a}{2a-1}$ Multiply by the reciprocal of the divisor.

$= \dfrac{5a(2a-1)}{11a^2(a+2)} \cdot \dfrac{a(a+2)}{2a-1}$ Factor

$= \dfrac{5}{11}$ Reduce.

29. $\dfrac{4x^2 - 4}{x^2 - 3x + 2} \div \dfrac{8x^2 + 32x + 24}{x^2 + 3x - 4}$

$= \dfrac{4x^2 - 4}{x^2 - 3x + 2} \cdot \dfrac{x^2 + 3x - 4}{8x^2 + 32x + 24}$ Multiply by the reciprocal of the divisor.

$= \dfrac{4(x + 1)(x - 1)}{(x - 2)(x - 1)} \cdot \dfrac{(x + 4)(x - 1)}{8(x + 3)(x + 1)}$ Factor.

$= \dfrac{1}{x - 2} \cdot \dfrac{(x + 4)(x - 1)}{2(x + 3)}$ Reduce.

$= \dfrac{(x + 4)(x - 1)}{2(x - 2)(x + 3)}$ Multiply.

33. $\dfrac{-28xyz}{r - t} \div \dfrac{2r + t}{-2y^2z}$

$= \dfrac{-28xyz}{r - t} \cdot \dfrac{-2y^2z}{2r + t}$ Multiply by the reciprocal of the divisor.

$= \dfrac{56xy^3z^2}{2r^2 - rt - t^2}$ Multiply.

37. $\dfrac{am}{2a + 4m} \div \dfrac{3}{-7m}$

$= \dfrac{am}{2a + 4m} \cdot \dfrac{-7m}{3}$ Multiply by the reciprocal of the divisor.

$= \dfrac{-7am^2}{6a + 12m}$ Multiply.

41. $\dfrac{6a - 7y}{9z + 7} \div \dfrac{4z - 8}{2a + 5y}$

$= \dfrac{6a - 7y}{9z + 7} \cdot \dfrac{2a + 5y}{4z - 8}$ Multiply by the reciprocal of the divisor.

$= \dfrac{12a^2 + 16ay - 35y^2}{36z^2 - 44z - 56}$ Multiply.

45. $\dfrac{5a^2 + 34a - 7}{2a^2 + 3a - 2} \div \dfrac{a^3 + 5a^2 - 14a}{a^3 - 4a}$

$= \dfrac{5a^2 + 34a - 7}{2a^2 + 3a - 2} \cdot \dfrac{a^3 - 4a}{a^3 + 5a^2 - 14a}$ Multiply by the reciprocal of the divisor.

$= \dfrac{(5a - 1)(a + 7)}{(2a - 1)(a + 2)} \cdot \dfrac{a(a + 2)(a - 2)}{a(a + 7)(a - 2)}$ Factor.

Exercises 5.2   127

$$= \frac{5a-1}{2a-1} \cdot \frac{1}{1}$$ Reduce.

$$= \frac{5a-1}{2a-1}$$ Multiply.

49. $\dfrac{3x^2 - 22x - 45}{2x^2 - 9x - 56} \div \dfrac{x^2 - 14x + 49}{2x^2 - 7x - 49}$

$$= \frac{3x^2 - 22x - 45}{2x^2 - 9x - 56} \cdot \frac{2x^2 - 7x - 49}{x^2 - 14x + 49}$$ Multiply by the reciprocal of the divisor.

$$= \frac{(3x+5)(x-9)}{(2x+7)(x-8)} \cdot \frac{(2x+7)(x-7)}{(x-7)^2}$$ Factor.

$$= \frac{(3x+5)(x-9)}{x-8} \cdot \frac{1}{x-7}$$ Reduce.

$$= \frac{(3x+5)(x-9)}{(x-8)(x-7)}$$ Multiply.

53. $r = V \div At$ Formula.

$r = \dfrac{V}{At}$ Rewrite as a fraction.

$r = \dfrac{V}{\pi r^2 t}$ $A = \pi r^2$.

$r = \dfrac{6}{(3.14)\left(\dfrac{1}{4}\right)^2 (60)}$ Replace V with 6, π with 3.14, r with $\dfrac{6}{12} = \dfrac{1}{4}$ and t with 1 min = 60 sec.

$r = 0.5096$ Simplify.

0.5096 ft per sec

57. $\dfrac{x^2 - 10x + 25}{10x^2 + 13x - 3} \cdot \dfrac{6 + x - 2x^2}{5x^2 - 24x - 5} \div \dfrac{x^2 - 7x + 10}{1 - 25x^2}$

$$= \frac{x^2 - 10x + 25}{10x^2 + 13x - 3} \cdot \frac{6 + x - 2x^2}{5x^2 - 24x - 5} \cdot \frac{1 - 25x^2}{x^2 - 7x + 10}$$ Change the division to multiplication.

$$= \frac{(x-5)^2}{(5x-1)(2x+3)} \cdot \frac{-1(2x+3)(x-2)}{(5x+1)(x-5)} \cdot \frac{-1(5x+1)(5x-1)}{(x-5)(x-2)}$$ Factor.

$$= \frac{1}{1} \cdot \frac{-1}{1} \cdot \frac{-1}{1}$$ Reduce.

$= 1$ Multiply.

65. $i = prt$ Formula.
$3600 = (10000)(0.12)t$ Replace i with 3600, p with 10000 and r with 0.12.
$3600 = 1200t$ Multiply.
$3 = t$ Divide both sides by 1200.

69. $7x^2 - 7x + 25x - 25$
    $= (7x^2 - 7x) + (25x - 25)$ Group the first two terms and the last two terms.
    $= 7x(x - 1) + 25(x - 1)$ Factor $7x$ out of the first group and 25 out of the second group.
    $= (x - 1)(7x + 25)$ Factor out $x - 1$.

EXERCISES 5.3

1. $3a = 3 \cdot a$ Completely factor each polynomial.
    $5a = 5 \cdot a$

    LCM $= 3 \cdot 5 \cdot a = 15a$ Write the product of the largest power of each factor.

5. $6x = 2 \cdot 3 \cdot x$ Completely factor each polynomial.
    $4x^2 = 2^2 \cdot x^2$
    $8 = 2^3$

    LCM $= 2^3 \cdot 3 \cdot x^2 = 24x^2$ Write the product of the largest power of each factor.

9. $3ab = 3 \cdot a \cdot b$ Completely factor each polynomial.
    $4a^2 = 2^2 \cdot a^2$
    $6b^2 = 2 \cdot 3 \cdot b^2$

    LCM $= 2^2 \cdot 3 \cdot a^2 \cdot b^2 = 12a^2b^2$ Write the product of the largest power of each factor.

13. $\dfrac{1}{a} + \dfrac{1}{4a} = 1$

    $4a\left(\dfrac{1}{a} + \dfrac{1}{4a}\right) = 4a(1)$ The LCM of the denominators is $4a$.
                                                            Multiply both sides by $4a$ to eliminate the fractions.

    $4a\left(\dfrac{1}{a}\right) + 4a\left(\dfrac{1}{4a}\right) = 4a(1)$ Distributive property.

    $4 + 1 = 4a$ Reduce and multiply.
    $5 = 4a$

    $\dfrac{5}{4} = a$

    $\left\{\dfrac{5}{4}\right\}$

17. $2ab = 2 \cdot a \cdot b$ Completely factor each polynomial.
    $4a^2 = 2^2 \cdot a^2$
    $8b^2 = 2^3 \cdot b^2$

    LCM $= 2^3 \cdot a^2 \cdot b^2$ Write the product of the largest power of each factor.
         $= 8a^2b^2$

21. $x - y = x - y$ Completely factor each polynomial.
    $x^2 - y^2 = (x - y)(x + y)$
    LCM $= (x - y)(x + y)$ Write the product of the largest power of each factor.

25. $2b^2 - 3b = b(2b - 3)$ Completely factor each polynomial.
    $10b^2 - 15b = 5b(2b - 3)$
    LCM $= 5b(2b - 3)$ Write the product of the largest power of each factor.

Exercises 5.3

29.
$$\frac{1}{x-2} + \frac{3}{x-2} = \frac{2}{3}$$

$$3(x-2)\left[\frac{1}{x-2} + \frac{3}{x-2}\right] = 3(x-2)\left(\frac{2}{3}\right)$$

The LCM of the denominators is $3(x-2)$. Multiply both sides by $3(x-2)$ to eliminate the fractions.

$$3(x-2)\left(\frac{1}{x-2}\right) + 3(x-2)\left(\frac{3}{x-2}\right) = 3(x-2)\left(\frac{2}{3}\right)$$

Distributive property.

$$3 + 9 = 2x - 4$$
$$12 = 2x - 4$$
$$16 = 2x$$
$$8 = x$$

Reduce and multiply.

{8}

33.
$$\frac{-6}{b-4} + \frac{2}{b-4} = \frac{4}{3}$$

$$3(b-4)\left[\frac{-6}{b-4} + \frac{2}{b-4}\right] = 3(b-4)\left(\frac{4}{3}\right)$$

The LCM of the denominators is $3(b-4)$. Multiply both sides by $3(b-4)$ to eliminate the fractions.

$$3(b-4)\left(\frac{-6}{b-4}\right) + 3(b-4)\left(\frac{2}{b-4}\right) = 3(b-4)\left(\frac{4}{3}\right)$$

Distributive property.

$$-18 + 6 = 4b - 16$$
$$-12 = 4b - 16$$
$$4 = 4b$$
$$1 = b$$

Reduce and multiply.

{1}

37. $b + c = b + c$
$b - c = b - c$
$b^2 - c^2 = (b + c)(b - c)$

Completely factor each polynomial.

LCM $= (b + c)(b - c)$

Write the product of the largest power of each factor.

41. $a^2 - b^2 = (a + b)(a - b)$
$3a - 3b = 3(a - b)$
$a^2 - 2ab + b^2 = (a - b)^2$

Completely factor each polynomial.

LCM $= 3(a + b)(a - b)^2$

Write the product of the largest power of each factor.

45.
$$\frac{4}{a-1} + \frac{2}{1-a} = \frac{3}{5}$$

$$\frac{4}{a-1} + \frac{2}{-(a-1)} = \frac{3}{5}$$

$1 - a = -(a - 1)$.

$$\frac{4}{a-1} - \frac{2}{a-1} = \frac{3}{5}$$

Combine signs.

$$5(a-1)\left[\frac{4}{a-1} - \frac{2}{a-1}\right] = 5(a-1)\left(\frac{3}{5}\right)$$

The LCM of the denominators is $5(a-1)$. Multiply both sides by $5(a-1)$ to eliminate the fractions.

$$5(a-1)\left(\frac{4}{a-1}\right) - 5(a-1)\left(\frac{2}{a-1}\right) = 5(a-1)\left(\frac{3}{5}\right)$$

Distributive property.

$$20 - 10 = 3a - 3$$
$$10 = 3a - 3$$
$$13 = 3a$$
$$\frac{13}{3} = a$$
$$\left\{\frac{13}{3}\right\}$$

Reduce and multiply.

49. $\dfrac{10}{x + 4} - \dfrac{3}{x - 2} = 0$

$(x + 4)(x - 2)\left[\dfrac{10}{x + 4} - \dfrac{3}{x - 2}\right] = (x + 4)(x - 2)(0)$

The LCM of the denominators is $(x + 4)(x - 2)$. Multiply both sides by $(x + 4)(x - 2)$ to eliminate the fractions.

$(x + 4)(x - 2)\left(\dfrac{10}{x + 4}\right) - (x + 4)(x - 2)\left(\dfrac{3}{x - 2}\right) = 0$

Distributive property. On the right side a factor of 0 results in a product of 0.

$$10(x - 2) - 3(x + 4) = 0 \quad \text{Reduce}$$
$$10x - 20 - 3x - 12 = 0 \quad \text{Multiply.}$$
$$7x - 32 = 0$$
$$7x = 32$$
$$x = \frac{32}{7}$$

$\left\{\dfrac{32}{7}\right\}$

53. $\dfrac{1}{x} + 1 + \dfrac{1}{2x + 2} = \dfrac{1}{x + 2}$

$\dfrac{1}{x + 1} + \dfrac{1}{2(x + 1)} = \dfrac{1}{x + 2}$

Factor the denominators.

$2(x + 1)(x + 2)\left[\dfrac{1}{x + 1} + \dfrac{1}{2(x + 1)}\right]$
$= 2(x + 1)(x + 2)\left(\dfrac{1}{x + 2}\right)$

The LCM of the denominators is $2(x + 1)(x + 2)$. Multiply both sides by $2(x + 1)(x + 2)$ to eliminate the fractions.

$2(x + 1)(x + 2)\left(\dfrac{1}{x + 1}\right) + 2(x + 1)(x + 2)\left[\dfrac{1}{2(x + 1)}\right]$
$= 2(x + 1)(x + 2)\left(\dfrac{1}{x + 2}\right)$

Distributive property.

$$2(x + 2) + x + 2 = 2(x + 1) \quad \text{Reduce.}$$
$$2x + 4 + x + 2 = 2x + 2 \quad \text{Multiply.}$$
$$3x + 6 = 2x + 2$$
$$x = -4$$

$\{-4\}$

57. $\dfrac{3}{x - 1} - \dfrac{4}{x + 1} = \dfrac{1}{x - 1}$

$(x - 1)(x + 1)\left[\dfrac{3}{x - 1} - \dfrac{4}{x + 1}\right] = (x - 1)(x + 1)\left(\dfrac{1}{x - 1}\right)$

The LCM of the denominators is $(x - 1)(x + 1)$. Multiply both sides by $(x - 1)(x + 1)$ to eliminate the fractions.

Exercises 5.3

$$(x - 1)(x + 1)\left(\frac{3}{x - 1}\right) - (x - 1)(x + 1)\left(\frac{4}{x + 1}\right)$$ Distributive property.

$$= (x - 1)(x + 1)\left(\frac{1}{x - 1}\right)$$

$$3(x + 1) - 4(x - 1) = x + 1$$ Reduce.
$$3x + 3 - 4x + 4 = x + 1$$ Multiply.
$$-x + 7 = x + 1$$
$$6 = 2x$$
$$3 = x$$

{3}

61. usual average speed: x
    twice as fast: $2x$

    $D = rt$ Formula.
    $594 = (2x)(11)$ Replace D with 594, r with $2x$ and t with 11.
    $594 = 22x$
    $27 = x$

    27 mph

65.

	rate	distance	time = distance/rate
express	$r + 10$	200	$200/(r + 10)$
regular	r	200	$200/r$

$$\begin{pmatrix}\text{time of}\\\text{express bus}\end{pmatrix} = \begin{pmatrix}\text{time of}\\\text{regular bus}\end{pmatrix} - \begin{pmatrix}\text{one}\\\text{hour}\end{pmatrix}$$

$$\frac{200}{r + 10} = \frac{200}{r} - 1$$ Substitute information from chart.

$$r(r + 10)\left(\frac{200}{r + 10}\right) = r(r + 10)\left[\frac{200}{r} - 1\right]$$ The LCM of the denominators is $r(r + 10)$. Multiply both sides by $r(r + 10)$ to eliminate the fractions.

$$r(r + 10)\left(\frac{200}{r + 10}\right) = r(r + 10)\left(\frac{200}{r}\right) - r(r + 10)(1)$$ Distributive property.

$$200r = 200r + 2000 - r^2 - 10r$$ Reduce and multiply.
$$r^2 + 10r - 2000 = 0$$
$$(r + 50)(r - 40) = 0$$
$$r + 50 = 0 \quad \text{or} \quad r - 40 = 0$$
$$r = -50 \quad \text{or} \quad r = 40$$

Since $r > 0$, $r = 40$ and $r + 10 = 40 + 10 = 50$.

Express bus: 50 mph; Regular bus: 40 mph

69. One possible answer: First completely factor all of the polynomials. Then the LCM is the product of the largest power of each factor.

73. $\dfrac{4x + 12}{6x^2} \cdot \dfrac{9x}{12x + 36} \div \dfrac{14x^2}{2x^2 + 10x} \div \dfrac{21x^2}{x + 5}$

$= \dfrac{4x + 12}{6x^2} \cdot \dfrac{9x}{12x + 36} \cdot \dfrac{2x^2 + 10x}{14x^2} \cdot \dfrac{x + 5}{21x^2}$ Change the division to multiplication.

$= \dfrac{4(x + 3)}{6x^2} \cdot \dfrac{9x}{12(x + 3)} \cdot \dfrac{2x(x + 5)}{14x^2} \cdot \dfrac{x + 5}{21x^2}$ Factor.

$= \dfrac{1}{1} \cdot \dfrac{1}{1} \cdot \dfrac{x + 5}{14} \cdot \dfrac{x + 5}{21x^2}$ Reduce.

$= \dfrac{(x + 5)^2}{249x^2}$ Multiply.

77. $(14a - 3b + 2c) - (3a - 4b) + (-2b - 3c)$
$= (14a - 3b + 2c) + (-3a + 4b) + (-2b - 3c)$ Change the subtraction to addition.
$= (14 - 3)a + (-3 + 4 - 2)b + (2 - 3)c$ Combine like terms.
$= 11a - b - c$

81. $x^2 - x - 56 = (x - 8)(x + 7)$

$mn = -56; \quad m + n = -1.$
$(-8)(7) = -56; \quad -8 + 7 = -1.$

85. $x^2 + 10x - 56 = (x + 14)(x - 4)$

$mn = -56; \quad m + n = 10.$
$(14)(-4) = -56; \quad 14 + (-4) = 10.$

EXERCISES 5.4

1. $\dfrac{6}{3 + x} - \dfrac{4}{3 + x} = \dfrac{6 - 4}{3 + x}$ Find the difference of the numerators.

$= \dfrac{2}{3 + x}$

5. $\dfrac{x}{y - 4} + \dfrac{z}{4 - y} = \dfrac{x}{y - 4} + \dfrac{z}{-(y - 4)}$ $4 - y = -(y - 4).$

$= \dfrac{x}{y - 4} + \dfrac{-z}{y - 4}$ $\dfrac{a}{-b} = \dfrac{-a}{b}.$

$= \dfrac{x + (-z)}{y - 4}$ Add the numerators.

$= \dfrac{x - z}{y - 4}$ Simplify.

9. $\dfrac{m}{2y} - \dfrac{3m}{8y} = \dfrac{m}{2y} \cdot \dfrac{4}{4} - \dfrac{3m}{8y}$ The LCM of the denominators is $8y$. Build each expression to have $8y$ as the common denominator.

$= \dfrac{4m}{8y} - \dfrac{3m}{8y}$

$= \dfrac{4m - 3m}{8y}$ Find the difference of the numerators.

$= \dfrac{m}{8y}$

Exercises 5.4

13. $\dfrac{2}{3a} + \dfrac{5}{6a} = \dfrac{2}{3a} \cdot \dfrac{2}{2} + \dfrac{5}{6a}$ The LCM of the denominators is $6a$. Build each expression to have $6a$ as the common denominator.

$= \dfrac{4}{6a} + \dfrac{5}{6a}$

$= \dfrac{4 + 5}{6a}$ Add the numerators.

$= \dfrac{9}{6a}$

$= \dfrac{3}{2a}$ Reduce.

17. $\dfrac{4}{x} + \dfrac{7}{y} = \dfrac{4}{x} \cdot \dfrac{y}{y} + \dfrac{7}{y} \cdot \dfrac{x}{x}$ The LCM of the denominators is xy. Build each expression to have xy as the common denominator.

$= \dfrac{4y}{xy} + \dfrac{7x}{xy}$

$= \dfrac{4y + 7x}{xy}$ Add the numerators.

21. $\dfrac{5x}{a - 4b} + \dfrac{y}{4b - a} = \dfrac{5x}{a - 4b} + \dfrac{y}{-(a - 4b)}$ $4b - a = -(a - 4b)$.

$= \dfrac{5x}{a - 4b} + \dfrac{-y}{a - 4b}$ $\dfrac{a}{-b} = \dfrac{-b}{a}$.

$= \dfrac{5x + (-y)}{a - 4b}$ Add the numerators.

$= \dfrac{5x - y}{a - 4b}$ Simplify.

25. $\dfrac{4}{y} + \dfrac{3}{y + 1} = \dfrac{4}{y} \cdot \dfrac{y + 1}{y + 1} + \dfrac{3}{y + 1} \cdot \dfrac{y}{y}$ The LCM of the denominators is $y(y + 1)$. Build each expression to have $y(y + 1)$ as the common denominator.

$= \dfrac{4y + 4}{y(y + 1)} + \dfrac{3y}{y(y + 1)}$

$= \dfrac{4y + 4 + 3y}{y(y + 1)}$ Add.

$= \dfrac{7y + 4}{y(y + 1)}$

29. $\dfrac{4}{a + b} - \dfrac{7}{a - b}$ The LCM of the denominators is $(a + b)(a - b)$. Build each expression to have $(a + b)(a - b)$ as the common denominator.

$= \dfrac{4}{a + b} \cdot \dfrac{a - b}{a - b} - \dfrac{7}{a - b} \cdot \dfrac{a + b}{a + b}$

$= \dfrac{4a - 4b}{(a + b)(a - b)} - \dfrac{7a + 7b}{(a + b)(a - b)}$

$$= \frac{4a - 4b - (7a + 7b)}{(a + b)(a - b)}$$ Subtract.

$$= \frac{4a - 4b - 7a - 7b}{(a + b)(a - b)}$$ Add the opposite.

$$= \frac{-3a - 11b}{(a + b)(a - b)}$$

33. $\dfrac{6}{7a} - \dfrac{3}{2a} + \dfrac{8}{a}$

$$= \frac{6}{7a} \cdot \frac{2}{2} - \frac{3}{2a} \cdot \frac{7}{7} + \frac{8}{a} \cdot \frac{14}{14}$$ The LCM of the denominators is $14a$. Build each expression to have $14a$ as the common denominator.

$$= \frac{12}{14a} - \frac{21}{14a} + \frac{112}{14a}$$

$$= \frac{12 - 21 + 112}{14a}$$ Add and subtract.

$$= \frac{103}{14a}$$

37. $\dfrac{-18}{b} + \dfrac{2}{a} - \dfrac{1}{c}$

$$= \frac{-18}{b} \cdot \frac{ac}{ac} + \frac{2}{a} \cdot \frac{bc}{bc} - \frac{1}{c} \cdot \frac{ab}{ab}$$ The LCM of the denominators is abc. Build each expression to have abc as the common denominator.

$$= \frac{-18ac}{abc} + \frac{2bc}{abc} - \frac{ab}{abc}$$

$$= \frac{-18ac + 2bc - ab}{abc}$$ Add and subtract.

41. $\dfrac{3x}{(x + 7)(x - 3)} - \dfrac{9}{(x - 3)(x + 7)}$

$$= \frac{3x - 9}{(x + 7)(x - 3)}$$ Find the difference of the numerators.

$$= \frac{3(x - 3)}{(x + 7)(x - 3)}$$ Factor the numerator.

$$= \frac{3}{x + 7}$$ Reduce.

45. $\dfrac{2}{x + 5} - \dfrac{7}{x - 1}$

$$= \frac{2}{x + 5} \cdot \frac{x - 1}{x - 1} - \frac{7}{x - 1} \cdot \frac{x + 5}{x + 5}$$ The LCM of the denominators is $(x + 5)(x - 1)$. Build each expression to have $(x + 5)(x - 1)$ as the common denominator.

$$= \frac{2x - 2}{(x + 5)(x - 1)} - \frac{7x + 35}{(x + 5)(x - 1)}$$

Exercises 5.4

$$= \frac{2x - 2 - (7x + 35)}{(x + 5)(x - 1)} \qquad \text{Subtract.}$$

$$= \frac{2x - 2 - 7x - 35}{(x + 5)(x - 1)} \qquad \text{Add the opposite.}$$

$$= \frac{-5x - 37}{(x + 5)(x - 1)}$$

49. $\dfrac{1}{x} + x - 3$

$$= \frac{1}{x} + \frac{x}{1} \cdot \frac{x}{x} - \frac{3}{1} \cdot \frac{x}{x} \qquad \text{Write } x \text{ as } \frac{x}{1} \text{ and } 3 \text{ as } \frac{3}{1}.\text{ Build each fraction to have the common denominator } x.$$

$$= \frac{1}{x} + \frac{x^2}{x} - \frac{3x}{x}$$

$$= \frac{1 + x^2 - 3x}{x} \qquad \text{Add and subtract.}$$

$$= \frac{x^2 - 3x + 1}{x}$$

53. $\dfrac{3}{(x + 4)(x - 5)} - \dfrac{4}{(x + 4)(5 - x)}$

$$= \frac{3}{(x + 4)(x - 5)} - \frac{4}{(x + 4)(-1)(x - 5)} \qquad 5 - x = -(x - 5).$$

$$= \frac{3}{(x + 4)(x - 5)} - \frac{-4}{(x + 4)(x - 5)} \qquad \frac{a}{-b} = \frac{-a}{b}.$$

$$= \frac{3 - (-4)}{(x + 4)(x - 5)} \qquad \text{Subtract.}$$

$$= \frac{7}{(x + 4)(x - 5)} \qquad \text{Simplify.}$$

57. $\dfrac{x + 1}{x - 2} + \dfrac{3}{2 - x} - \dfrac{4 - x}{x}$

$$= \frac{x + 1}{x - 2} + \frac{3}{-(x - 2)} - \frac{4 - x}{x} \qquad 2 - x = -(x - 2).$$

$$= \frac{x + 1}{x - 2} + \frac{-3}{x - 2} - \frac{4 - x}{x} \qquad \frac{a}{-b} = \frac{-a}{b}.$$

$$= \frac{x + 1}{x - 2} \cdot \frac{x}{x} + \frac{-3}{x - 2} \cdot \frac{x}{x} - \frac{4 - x}{x} \cdot \frac{x - 2}{x - 2} \qquad \text{Build each fraction to have the common denominator } x(x - 2).$$

$$= \frac{x^2 + x}{x(x - 2)} + \frac{-3x}{x(x - 2)} - \frac{-x^2 + 6x - 8}{x(x - 2)}$$

$$= \frac{x^2 + x + (-3x) - (-x^2 + 6x - 8)}{x(x - 2)} \qquad \text{Add and subtract.}$$

$$= \frac{x^2 + x - 3x + x^2 - 6x + 8}{x(x - 2)}$$ Simplify.

$$= \frac{2x^2 - 8x + 8}{x(x - 2)}$$

$$= \frac{2(x - 2)^2}{x(x - 2)}$$ Factor the numerator.

$$= \frac{2(x - 2)}{x}$$ Reduce.

$$= \frac{2x - 4}{x}$$

61. $\dfrac{3}{y - 5} + \dfrac{4}{y + 6} - \dfrac{2y - 1}{y^2 + y - 30}$

$$= \frac{3}{y - 5} + \frac{4}{y + 6} - \frac{2y - 1}{(y - 5)(y + 6)}$$ $y^2 + y - 30 = (y - 5)(y + 6)$.

$$= \frac{3}{y - 5} \cdot \frac{y + 6}{y + 6} + \frac{4}{y + 6} \cdot \frac{y - 5}{y - 5} - \frac{2y - 1}{(y - 5)(y + 6)}$$ Build each fraction to have the common denominator $(y - 5)(y + 6)$.

$$= \frac{3y + 18}{(y - 5)(y + 6)} + \frac{4y - 20}{(y - 5)(y + 6)} - \frac{2y - 1}{(y - 5)(y + 6)}$$

$$= \frac{3y + 18 + 4y - 20 - (2y - 1)}{(y - 5)(y + 6)}$$ Add and subtract.

$$= \frac{3y + 18 + 4y - 20 - 2y + 1}{(y - 5)(y + 6)}$$ Simplify.

$$= \frac{5y - 1}{(y - 5)(y + 6)}$$

65. number: x

$$5\left(\frac{1}{3x}\right) - 3\left(\frac{1}{2x}\right) = \frac{1}{30}$$

$\dfrac{1}{3x}$: reciprocal of three times the number.

$\dfrac{1}{2x}$: reciprocal of twice the number.

$$\frac{5}{3x} - \frac{3}{2x} = \frac{1}{30}$$ Multiply.

$$30x\left(\frac{5}{3x} - \frac{3}{2x}\right) = 30x\left(\frac{1}{30}\right)$$ Multiply by the least common denominator, $30x$, to eliminate the fractions.

$$30x\left(\frac{5}{3x}\right) - 30x\left(\frac{3}{2x}\right) = 30x\left(\frac{1}{30}\right)$$ Distributive property.

$$50 - 45 = x$$ Reduce and multiply.
$$5 = x$$

The number is 5.

Exercises 5.4

69. $\dfrac{1}{R} = \dfrac{1}{r_1} + \dfrac{1}{r_2} + \dfrac{1}{r_3} + \dfrac{1}{r_4}$

$\dfrac{1}{R} = \dfrac{1}{2r} + \dfrac{1}{r-1} + \dfrac{1}{4r} + \dfrac{1}{2r-2}$ Replace r_1 with $2r$, r_2 with $r-1$, r_3 with $4r$ and r_4 with $2r-2$.

$\dfrac{1}{R} = \dfrac{1}{2r} + \dfrac{1}{r-1} + \dfrac{1}{4r} + \dfrac{1}{2(r-1)}$ $2r - 2 = 2(r-1)$.

$\dfrac{1}{R} = \dfrac{1}{2r} \cdot \dfrac{2(r-1)}{2(r-1)} + \dfrac{1}{r-1} \cdot \dfrac{4r}{4r} + \dfrac{1}{4r} \cdot \dfrac{r-1}{r-1}$ Build each fraction to the common denominator $4r(r-1)$.

$\qquad + \dfrac{1}{2(r-1)} \cdot \dfrac{2r}{2r}$

$\dfrac{1}{R} = \dfrac{2r-2}{4r(r-1)} + \dfrac{4r}{4r(r-1)} + \dfrac{r-1}{4r(r-1)} + \dfrac{2r}{4r(r-1)}$

$\dfrac{1}{R} = \dfrac{2r - 2 + 4r + r - 1 + 2r}{4r(r-1)}$ Add.

$\dfrac{1}{R} = \dfrac{9r - 3}{4r(r-1)}$

Let $r = 3$:

$\dfrac{1}{R} = \dfrac{9(3) - 3}{4(3)(3-1)}$ Replace r with 3.

$\dfrac{1}{R} = \dfrac{24}{24}$

$\dfrac{1}{R} = 1$

$R = 1$ ohm

Let $r = 4$:

$\dfrac{1}{R} = \dfrac{9(4) - 3}{4(4)(4-1)}$ Replace r with 4.

$\dfrac{1}{R} = \dfrac{33}{48}$

$R = \dfrac{48}{33} = \dfrac{16}{11} = 1\dfrac{5}{11}$ ohms

Let $r = 10$:

$\dfrac{1}{R} = \dfrac{9(10) - 3}{4(10)(10-1)}$ Replace r with 10.

$\dfrac{1}{R} = \dfrac{87}{360}$

$R = \dfrac{360}{87} = \dfrac{120}{29} = 4\dfrac{4}{29}$ ohms   Take the reciprocal of both sides and reduce.

73. One possible answer: Here we are finding the sum of an expression. We do not have an equation and hence cannot multiply by the least common denominator to eliminate the fractions.

$$\frac{2}{x^2 - 5x + 6} + \frac{3}{x^2 - x - 6}$$

$$= \frac{2}{(x - 2)(x - 3)} + \frac{3}{(x + 2)(x - 3)}$$

$$= \frac{2}{(x - 2)(x - 3)} \cdot \frac{x + 2}{x + 2} + \frac{3}{(x + 2)(x - 3)} \cdot \frac{x - 2}{x - 2}$$
Build each fraction to the common denominator $(x - 2)(x - 3)(x + 2)$.

$$= \frac{2x + 4}{(x - 2)(x - 3)(x + 2)} + \frac{3x - 6}{(x - 2)(x - 3)(x + 2)}$$
Add.

$$= \frac{2x + 4 + 3x - 6}{(x - 2)(x - 3)(x + 2)}$$

$$= \frac{5x - 2}{(x - 2)(x - 3)(x + 2)}$$

77. rate of river: x

	rate	distance	time = distance/rate
downstream	$20 + x$	60	$60/(20 + x)$
upstream	$20 - x$	40	$40/(20 - x)$

$$\begin{pmatrix} \text{time} \\ \text{downstream} \end{pmatrix} = \begin{pmatrix} \text{time} \\ \text{upstream} \end{pmatrix}$$

$$\frac{60}{20 + x} = \frac{40}{20 - x}$$
Substitute information from the chart.

$$(20 + x)(20 - x)\left(\frac{60}{20 + x}\right) = (20 + x)(20 - x)\left(\frac{40}{20 - x}\right)$$
Multiply both sides by the least common denominator $(20 + x)(20 - x)$ to eliminate the fractions.

$$1200 - 60x = 800 + 40x$$
Reduce and multiply.
$$400 = 100x$$
$$4 = x$$

4 mph

81. $2[y - 3(4 - 2y) + 6] - 3$
 $= 2(y - 12 + 6y + 6) - 3$ Multiply inside the brackets.
 $= 2(7y - 6) - 3$ Combine like terms inside the parentheses.
 $= 14y - 12 - 3$ Multiply.
 $= 14y - 15$ Combine like terms.

Exercises 5.5

85. $64x^2y^2 - 49z^2$
    $= (8xy)^2 - (7z)^2$ Write $64x^2y^2$ as the square of $8xy$ and $49z^2$ as the square of $7z$.
    $= (8xy + 7z)(8xy - 7z)$ To factor write the sum and difference of the terms that were squared.

EXERCISES 5.5

1. $\dfrac{\frac{2}{a}}{\frac{5}{b}} = \dfrac{2}{a} \div \dfrac{5}{b}$

    $= \dfrac{2}{a} \cdot \dfrac{b}{5}$ Multiply by the reciprocal of the divisor.

    $= \dfrac{2b}{5a}$ Multiply.

5. $\dfrac{\frac{1}{4}}{\frac{x}{y}} = \dfrac{1}{4} \div \dfrac{x}{y}$

    $= \dfrac{1}{4} \cdot \dfrac{y}{x}$ Multiply by reciprocal of the divisor.

    $= \dfrac{y}{4x}$ Multiply.

9. $\dfrac{\frac{x^2}{y^2}}{\frac{x}{y}} = \dfrac{x^2}{y^2} \div \dfrac{x}{y}$

    $= \dfrac{x^2}{y^2} \cdot \dfrac{y}{x}$ Multiply by the reciprocal of the divisor.

    $= \dfrac{x}{y}$ Reduce and multiply.

13. $\dfrac{\frac{5x^2y}{4ab}}{\frac{15xy}{8a^2b^2}} = \dfrac{5x^2y}{4ab} \div \dfrac{15xy}{8a^2b^2}$

    $= \dfrac{5x^2y}{4ab} \cdot \dfrac{8a^2b^2}{15xy}$ Multiply by the reciprocal of the divisor.

    $= \dfrac{x}{1} \cdot \dfrac{2ab}{3}$ Reduce.

    $= \dfrac{2abx}{3}$ Multiply.

17. $\dfrac{\dfrac{x+1}{x-1}}{\dfrac{x+1}{x}} = \dfrac{x+1}{x-1} \div \dfrac{x+1}{x}$

$= \dfrac{x+1}{x-1} \cdot \dfrac{x}{x+1}$ Multiply by the reciprocal of the divisor.

$= \dfrac{x}{x-1}$ Reduce and multiply.

21. $\dfrac{\dfrac{x^2-1}{x}}{\dfrac{x+1}{x}} = \dfrac{x^2-1}{x} \div \dfrac{x+1}{x}$

$= \dfrac{x^2-1}{x} \cdot \dfrac{x}{x+1}$ Multiply by the reciprocal of the divisor.

$= \dfrac{(x+1)(x-1)}{x} \cdot \dfrac{x}{x+1}$ Factor $x^2 - 1$.

$= x - 1$ Reduce and multiply.

25. $\dfrac{2 + \dfrac{3}{a}}{3 - \dfrac{1}{a}} = \dfrac{a\left(2 + \dfrac{3}{a}\right)}{a\left(3 - \dfrac{1}{a}\right)}$ Multiply both the numerator and the denominator by the LCM of the denominators, a.

$= \dfrac{2a + 3}{3a - 1}$ Simplify.

29. $\dfrac{2 + \dfrac{3}{2x+1}}{3 - \dfrac{1}{2x+1}} = \dfrac{(2x+1)\left(2 + \dfrac{3}{2x+1}\right)}{(2x+1)\left(3 - \dfrac{1}{2x+1}\right)}$ Multiply both the numerator and the denominator by the LCM of the denominators, $2x + 1$.

$= \dfrac{2(2x+1) + 3}{3(2x+1) - 1}$ Simplify.

$= \dfrac{4x + 2 + 3}{6x + 3 - 1}$

$= \dfrac{4x + 5}{6x + 2}$

33. $\dfrac{\dfrac{x}{5x+3} - 5}{\dfrac{-x}{5x+3} + 7} = \dfrac{(5x+3)\left(\dfrac{x}{5x+3} - 5\right)}{(5x+3)\left(\dfrac{-x}{5x+3} + 7\right)}$ Multiply by the LCM, $5x + 3$.

$= \dfrac{x - 5(5x+3)}{-x + 7(5x+3)}$ Simplify.

Exercises 5.5

$$= \frac{x - 25x - 15}{-x + 35x + 21}$$

$$= \frac{-24x - 15}{34x + 21}$$

37. $\dfrac{\dfrac{x}{y} - \dfrac{y}{x}}{\dfrac{x}{y} + \dfrac{y}{x}} = \dfrac{xy\left(\dfrac{x}{y} - \dfrac{y}{x}\right)}{xy\left(\dfrac{x}{y} + \dfrac{y}{x}\right)}$ Multiply by the LCM, xy.

$$= \frac{x^2 - y^2}{x^2 + y^2}$$ Simplify.

41. $\dfrac{\dfrac{b}{c} - \dfrac{c}{b}}{\dfrac{b}{c} + \dfrac{c}{b}} = \dfrac{bc\left(\dfrac{b}{c} - \dfrac{c}{b}\right)}{bc\left(\dfrac{b}{c} + \dfrac{c}{b}\right)}$ Multiply by the LCM, bc.

$$= \frac{b^2 - c^2}{b^2 + c^2}$$ Simplify.

45. $\dfrac{\dfrac{1}{x-1} + \dfrac{1}{x+1}}{\dfrac{1}{x-1} - \dfrac{1}{x+1}}$

$$= \frac{(x+1)(x-1)\left(\dfrac{1}{x-1} + \dfrac{1}{x+1}\right)}{(x+1)(x-1)\left(\dfrac{1}{x-1} - \dfrac{1}{x+1}\right)}$$ Multiply by the LCM, $(x+1)(x-1)$.

$$= \frac{x + 1 + x - 1}{x + 1 - (x - 1)}$$ Simplify.

$$= \frac{2x}{x + 1 - x + 1}$$ $-(x - 1) = -x + 1$.

$$= \frac{2x}{2}$$

$$= x$$ Reduce.

49. $\dfrac{\dfrac{4xy^2 - 8xy}{5a - 10a^2}}{\dfrac{5xy - 10x}{3a - 6a^2}}$

$$= \frac{4xy^2 - 8xy}{5a - 10a^2} \div \frac{5xy - 10x}{3a - 6a^2}$$

$$= \frac{4xy^2 - 8xy}{5a - 10a^2} \cdot \frac{3a - 6a^2}{5xy - 10x}$$ Multiply by the reciprocal of the divisor.

$$= \frac{4xy(y-2)}{5a(1-2a)} \cdot \frac{3a(1-2a)}{5x(y-2)} \qquad \text{Factor.}$$

$$= \frac{4y}{5} \cdot \frac{3}{5} \qquad \text{Reduce.}$$

$$= \frac{12y}{25} \qquad \text{Multiply.}$$

53.

	rate	distance	time = distance/rate
going	60	d	$d/60$
returning	30	d	$d/30$

$$\text{Average speed for round trip} = \frac{\text{Total distance}}{\text{Total time}}$$

$$= \frac{d+d}{\dfrac{d}{60} + \dfrac{d}{30}} \qquad \text{Substitute information from the chart.}$$

$$= \frac{2d}{\dfrac{d}{60} + \dfrac{d}{30}}$$

$$= \frac{60(2d)}{60\left(\dfrac{d}{60} + \dfrac{d}{30}\right)} \qquad \text{Multiply by the LCM, 60.}$$

$$= \frac{120d}{d + 2d} \qquad \text{Simplify.}$$

$$= \frac{120d}{3d}$$

$$= 40 \qquad \text{Reduce.}$$

40 mph

57. $v' = \dfrac{\dfrac{m_1}{v_2} + \dfrac{m_2}{v_1}}{\dfrac{m_1 + m_2}{v_1 v_2}}$

$$= \frac{v_1 v_2 \left(\dfrac{m_1}{v_2} + \dfrac{m_2}{v_1}\right)}{v_1 v_2 \left(\dfrac{m_1 + m_2}{v_1 v_2}\right)} \qquad \text{Multiply by the LCM, } v_1 v_2.$$

$$= \frac{m_1 v_1 + m_2 v_2}{m_1 + m_2} \qquad \text{Simplify.}$$

Exercises 5.5

61. One possible answer: In one method you multiply by the reciprocal of the divisor, reduce and simplify. In the other method you multiply the numerator and the denominator by the LCM of the denominators. This eliminates the fractions within the complex fraction. Again you multiply and simplify.

65. $\dfrac{3 - \dfrac{4}{x+1}}{\dfrac{4}{x^2-1} - \dfrac{3}{x-1}}$

$= \dfrac{3 - \dfrac{4}{x+1}}{\dfrac{4}{(x+1)(x-1)} - \dfrac{3}{x-1}}$ $x^2 - 1 = (x+1)(x-1)$.

$= \dfrac{(x+1)(x-1)\left(3 - \dfrac{4}{x+1}\right)}{(x+1)(x-1)\left[\dfrac{4}{(x+1)(x-1)} - \dfrac{3}{x-1}\right]}$ Multiply by the LCM, $(x+1)(x-1)$.

$= \dfrac{3(x^2-1) - 4(x-1)}{4 - 3(x+1)}$ Simplify.

$= \dfrac{3x^2 - 3 - 4x + 4}{4 - 3x - 3}$

$= \dfrac{3x^2 - 4x + 1}{-3x + 1}$

$= \dfrac{(3x-1)(x-1)}{-(3x-1)}$ Factor. $-3x + 1 = -(3x - 1)$.

$= \dfrac{x-1}{-1}$ Reduce.

$= 1 - x$ Simplify.

69. $(2x - 11)(x + 4) = 2x^2 + 8x - 11x - 44$ Multiply using FOIL.
$ = 2x^2 - 3x - 44$ Combine like terms.

73. $2x^2 + x - 15 = 2x^2 + 6x - 5x - 15$ $mn = 2(-15); \; m + n = 1.$
$ = -30$
$(6)(-5) = -30; \; 6 + (-5) = 1.$
Rewrite the trinomial.

$ = 2x(x+3) - 5(x+3)$ Factor by grouping.
$ = (x+3)(2x-5)$

77. $35x^2 - 11x - 72 = 35x^2 + 45x - 56x - 72$ $mn = 35(-72);$ $m + n = -11.$
$ = -2520$
$ = 5x(7x + 9) - 8(7x + 9)$ $(45)(-56) = -2520;$ $45 + (-56) = -11.$
$ = (7x + 9)(5x - 8)$ Rewrite the trinomial.
    Factor by grouping.

EXERCISES 5.6

1. $3 \text{ yards} = \dfrac{3 \text{ yds}}{1} \cdot \dfrac{36 \text{ inches}}{1 \text{ yd}}$ There are 36 inches in 1 yard.

$\phantom{3 \text{ yards}} = \dfrac{3 \cdot 36 \text{ inches}}{1}$ Eliminate the like measures.

$\phantom{3 \text{ yards}} = 108 \text{ inches}$

5. $1 \text{ m} = \dfrac{1 \text{ m}}{1} \cdot \dfrac{100 \text{ cm}}{1 \text{ m}}$ There are 100 cm in 1 m.

$\phantom{1 \text{ m}} = \dfrac{1 \cdot 100 \text{ cm}}{1}$ Eliminate the like measures.

$\phantom{1 \text{ m}} = 100 \text{ cm}$

9. $\dfrac{5}{2} = \dfrac{w}{8}$

$2w = 40$
$w = 20$ Cross multiply.

13. $\dfrac{11}{5} = \dfrac{y}{30}$

$5y = 330$
$y = 66$ Cross multiply.

17. $\dfrac{21}{x} = \dfrac{63}{12}$

$63x = 252$
$x = 4$ Cross multiply.

21. $1 \text{ ft}^2 = \dfrac{1 \text{ ft} \cdot \text{ft}}{1} \cdot \dfrac{12 \text{ in}}{1 \text{ ft}} \cdot \dfrac{12 \text{ in}}{1 \text{ ft}}$ There are 12 inches in 1 foot.

$\phantom{1 \text{ ft}^2} = \dfrac{1 \cdot 12 \cdot 12 \cdot \text{in} \cdot \text{in}}{1 \cdot 1 \cdot 1}$ Eliminate the like measures.

$\phantom{1 \text{ ft}^2} = 144 \text{ in}^2$

Exercises 5.6

25. $\dfrac{2 \text{ mi}}{1 \text{ hr}} = \dfrac{2 \text{ mi}}{1 \text{ hr}} \cdot \dfrac{5280 \text{ ft}}{1 \text{ mi}}$ There are 5280 ft in 1 mile.

$= \dfrac{2 \cdot 5280 \text{ ft}}{1 \text{ hr}}$ Eliminate the like measures.

$= \dfrac{10560 \text{ ft}}{1 \text{ hr}}$

29. $\dfrac{4}{5} = \dfrac{x}{\frac{1}{2}}$

$5x = 2$ Cross multiply.

$x = \dfrac{2}{5}$

33. $\dfrac{4}{\frac{2}{3}} = \dfrac{x}{1}$

$\dfrac{2}{3}x = 4$ Cross multiply.

$x = 6$

37. $\dfrac{\frac{3}{8}}{\frac{6}{5}} = \dfrac{x}{16}$

$\dfrac{6}{5}x = 6$ Cross multiply.

$x = 5$

41. $\dfrac{10 \text{ mi}}{1 \text{ hr}} = \dfrac{10 \text{ mi}}{1 \text{ hr}} \cdot \dfrac{5280 \text{ ft}}{1 \text{ mi}} \cdot \dfrac{1 \text{ hr}}{60 \text{ min}} \cdot \dfrac{1 \text{ min}}{60 \text{ sec}}$ There are 5280 feet in 1 mile, 60 minutes in 1 hour and 60 seconds in 1 minute.

$= \dfrac{10 \cdot 5280 \cdot 1 \cdot 1 \text{ ft}}{1 \cdot 1 \cdot 60 \cdot 60 \text{ sec}}$ Eliminate the like measures.

$= \dfrac{14\frac{2}{3} \text{ ft}}{1 \text{ sec}}$

45. $\dfrac{450 \text{ words}}{1 \text{ min}}$

$= \dfrac{450 \text{ words}}{1 \text{ min}} \cdot \dfrac{60 \text{ min}}{1 \text{ hr}} \cdot \dfrac{1 \text{ page}}{250 \text{ words}}$ There are 60 minutes in 1 hour.

$$= \frac{450 \cdot 60 \cdot 1 \text{ page}}{1 \cdot 1 \cdot 250 \text{ hr}}$$ There are 60 minutes in 1 hour.

$$= \frac{108 \text{ pages}}{1 \text{ hr}}$$ Eliminate the like measures.

49. $\dfrac{0.055}{1} = \dfrac{R}{100}$

$R = 5.5$ Cross multiply.

53. $\dfrac{12}{25} = \dfrac{2x}{14}$

$50x = 168$ Cross multiply.
$x = 3.36$

57. $\dfrac{235}{376} = \dfrac{c}{12}$

$376c = 2820$ Cross multiply.
$c = 7.5$

61. $\dfrac{\text{width}}{\text{height}}: \dfrac{8}{10} = \dfrac{x}{25}$ Set up ratios as width to height.

$10x = 200$ Cross multiply.
$x = 20$

20 cm

65. $\dfrac{64 \text{ lbs}}{1 \text{ ft}^3} = \dfrac{? \text{ oz}}{1 \text{ in}^3}$

$\dfrac{64 \text{ lbs}}{1 \text{ ft}^3} = \dfrac{64 \text{ lbs}}{1 \text{ ft} \cdot \text{ft} \cdot \text{ft}} \cdot \dfrac{16 \text{ oz}}{1 \text{ lb}} \cdot \dfrac{1 \text{ ft}}{12 \text{ in}} \cdot \dfrac{1 \text{ ft}}{12 \text{ in}} \cdot \dfrac{1 \text{ ft}}{12 \text{ in}}$ There are 16 ounces in 1 pound.
There are 12 inches in 1 foot.

$$= \frac{64 \cdot 16 \cdot 1 \cdot 1 \cdot 1 \text{ oz}}{1 \cdot 1 \cdot 12 \cdot 12 \cdot 12 \text{ in}^3}$$ Eliminate the like measures.

$$= \frac{0.6 \text{ oz}}{1 \text{ in}^3}$$

69. $\dfrac{\text{lb}}{\text{acre}}: \dfrac{100}{50} = \dfrac{x}{185}$ Set up ratios as pounds to acres.

$50x = 18500$ Cross multiply.
$x = 370$

370 lb

Exercises 5.6

73. $\dfrac{\text{games}}{\text{points}}$: $\dfrac{15}{1150} = \dfrac{9}{x}$

$15x = 10350$
$x = 690$

690 points

Set up ratios as games to points.
24 − 15 = 9 games remaining.
Cross multiply.

77. $\dfrac{\text{dollars}}{\text{tons}}$: $\dfrac{550}{\frac{3}{4}} = \dfrac{x}{2\frac{1}{4}}$

$\dfrac{3}{4}x = \dfrac{2475}{2}$

$x = 1650$

$1650

Set up ratios as dollars to tons.

Cross multiply.

81. One possible answer: It is correct. If $\dfrac{a}{b} = \dfrac{c}{d}$, then when we cross multiply we see that $ad = bc$. Divide both sides of this equation by ac.

$$\dfrac{ad}{ac} = \dfrac{bc}{ac}$$

$$\dfrac{d}{c} = \dfrac{b}{a}$$

85. $\dfrac{2x - 3}{x} = \dfrac{2}{x - 1}$

$2x = (2x - 3)(x - 1)$
$2x = 2x^2 - 5x + 3$
$0 = 2x^2 - 7x + 3$
$0 = (2x - 1)(x - 3)$
$2x - 1 = 0$ or $x - 3 = 0$
$2x = 1$ or $x = 3$
$x = \dfrac{1}{2}$

$\left\{\dfrac{1}{2}, 3\right\}$

Cross multiply.
Multiply using FOIL.
Subtract $2x$ from both sides.
Factor the right side.
Zero-product property.

89. $x^2 - 7x - 144 = (x - 16)(x + 9)$

$mn = -144;\ m + n = -7$.
$(-16)(9) = -144;\ -16 + 9 = -7$.

93. $25x^2 - 36y^2 = (5x)^2 - (6y)^2$
$= (5x + 6y)(5x - 6y)$

Write $25x^2$ as the square of $5x$ and $36y^2$ as the square of $6y$.
To factor, write the sum and difference of the terms that were squared.

97. $3x^2 - 4x = 15$
$3x^2 - 4x - 15 = 0$
$(3x + 5)(x - 3) = 0$

Subtract 15 from both sides.
Factor the left side.

$$3x + 5 = 0 \quad \text{or} \quad x - 3 = 0 \qquad \text{Zero-product property.}$$
$$3x = -5 \quad \text{or} \quad x = 3$$
$$x = -\frac{5}{3}$$
$$\left\{-\frac{5}{3}, 3\right\}$$

EXERCISES 5.7

1. $d = \dfrac{k}{r}$ Equation of inverse variation.

5. $V = kBh$ Equation of joint variation.

9. $I = kE$ Equation of direct variation.

13. $y = kx$ Equation of direct variation.
    $12 = k(8)$ Replace y with 12 and x with 8.
    $\dfrac{3}{2} = k$ Solve for k.

17. $y = kx^2$ Equation of direct variation.
    $14 = k(7)^2$ Replace y with 14 and x with 7.
    $14 = 49k$
    $\dfrac{2}{7} = k$ Solve for k.

21. $y = \dfrac{k}{x^3}$ Equation of inverse variation.
    $6 = \dfrac{k}{(2)^3}$ Replace y with 6 and x with 2.
    $6 = \dfrac{k}{8}$
    $48 = k$ Solve for k.

25. $y = kxz$ Equation of joint variation.
    $12 = k(4)(12)$ Replace y with 12, x with 4 and z with 12.
    $12 = 48k$
    $\dfrac{1}{4} = k$ Solve for k.

29. $a = kb$ Equation of direct variation.
    $14 = k(6)$ Replace a with 14 and b with 6.
    $\dfrac{7}{3} = k$ Solve for k.
    $a = \dfrac{7}{3}b$ Equation of variation, substituting $\dfrac{7}{3}$ for k.

Exercises 5.7

$a = \frac{7}{3}(16)$ Replace b with 16.

$a = \frac{112}{3}$ Solve for a.

33. $m = kn^2$ Equation of direct variation.
     $30 = k(5)^2$ Replace m with 30 and n with 5.
     $30 = 25k$

     $\frac{6}{5} = k$ Solve for k.

     $m = \frac{6}{5}n^2$ Equation of variation, substituting $\frac{6}{5}$ for k.

     $m = \frac{6}{5}(16)^2$ Replace n with 16.

     $m = 307.2$ Solve for m.

37. $z = kw^3$ Equation of direct variation.
     $81 = k(3)^3$ Replace z with 81 and w with 3.
     $81 = 27k$
     $3 = k$ Solve for k.
     $z = 3w^3$ Equation of variation, substituting 3 for k.
     $z = 3(2)^3$ Replace w with 2.
     $z = 24$ Solve for z.

41. $y = kxz^2$ Equation of joint variation.
     $36 = k(12)(3)^2$ Replace y with 36, x with 12 and z with 3.
     $36 = 108k$

     $\frac{1}{3} = k$ Solve for k.

     $y = \frac{1}{3}xz^2$ Equation of variation substituting $\frac{1}{3}$ for k.

     $y = \frac{1}{3}(8)(6)^2$ Replace x with 8 and z with 6.

     $y = 96$ Solve for y.

45. $p = ks$ Let p represent salary and s represent sales.
     $372 = k(3100)$ Replace p with 372 and s with 3100.

     $\frac{3}{25} = k$ Solve for k.

     $p = \frac{3}{25}s$ Equation of variation substituting $\frac{3}{25}$ for k.

     $p = \frac{3}{25}(4500)$ Replace s with 4500.

     $p = 540$ Solve for p.

     $540

49. $l = \dfrac{k}{w}$ Equation of inverse variation.

$12 = \dfrac{k}{8}$ Replace l with 12 and w with 8.

$96 = k$ Solve for k.

$l = \dfrac{96}{w}$ Equation of variation substituting 96 for k.

$l = \dfrac{96}{6}$ Replace w with 6.

$l = 16$ Solve for l.

16 ft

53. One possible answer: Since a varies directly with b, $a = kb$.

Double b: $a = k(2b) = 2kb = 2a$
a is doubled.

Triple b: $a = k(3b) = 3kb = 3a$
a is tripled.

Multiplied by 10: $a = k(10b) - 10kb = 10a$
a is multiplied by 10.

Divided by 4: $a = k\left(\dfrac{b}{4}\right) = \dfrac{1}{4}kb = \dfrac{1}{4}a$
a is divided by 4.

57. $F = \dfrac{k}{r^2}$ Equation of inverse variation.

$F = \dfrac{k}{(3r)^2}$ Triple the distance.

$= \dfrac{k}{9r^2}$

$= \dfrac{1}{9} \cdot \dfrac{k}{r^2}$

$= \dfrac{1}{9}F$ Replace $\dfrac{k}{r^2}$ with F.

The force is $\dfrac{1}{9}$ as strong.

61. $25x^2 - 100 = 25(x^2 - 4)$ The GCF is 25.
$= 25(x^2 - 2^2)$ Write 4 as 2^2.
$= 25(x + 2)(x - 2)$ Factor the difference of two squares.

65. $\dfrac{14m}{5n} \cdot \dfrac{7m}{13n} = \dfrac{98m^2}{65n^2}$ There are no common factors, so multiply the numerators and multiply the denominators.

69. $\dfrac{4c - 5d}{c + 2d} \div \dfrac{2c - d}{4c}$

$= \dfrac{4c - 5d}{c + 2d} \cdot \dfrac{4c}{2c - d}$ Multiply by the reciprocal of the divisor.

$= \dfrac{4c(4c - 5d)}{(c + 2d)(2c - d)}$ Multiply.

$= \dfrac{16c^2 - 20cd}{2c^2 + 3cd - 2d^2}$

CHAPTER 5 CONCEPT REVIEW

1. True

2. False; It is not a real number when x is replaced by 3.

3. True

4. True

5. False; Multiply by the reciprocal of the divisor.

6. False; It is not true when $x = 0$ or $x = 2$.

7. False; The value is not changed.

8. True

9. False; The sum is found by writing the fractions with a common denominator and then adding the numerators and keeping the common denominator.

10. True

11. True

12. False; First write with common denominators.

13. False; Also used to check for the value that makes the denominator zero.

14. True

15. False; It is also used to solve equations containing rational expressions.

16. True

17. True

18. True

19. False; Not the difference, the quotient.

20. True

21. True

22. True

23. True

24. False; The other decreases in the same ratio.

25. True

CHAPTER 5 TEST

1. $\dfrac{14}{8x} = \dfrac{?}{32x^2y}$

   $\dfrac{14}{8x} \cdot \dfrac{4xy}{4xy} = \dfrac{56xy}{32x^2y}$ $8x \cdot 4xy = 32x^2y$.
   Multiply both the numerator and the denominator by $4xy$.

2. $\dfrac{-5w}{w+10} - \dfrac{2w}{w+10}$

   $= \dfrac{-5w - 2w}{w+10}$ Subtract the numerators.

   $= \dfrac{-7w}{w+10}$

3. $\dfrac{15}{16} = \dfrac{y}{8}$

   $16y = 120$ Cross multiply.
   $y = 7.5$

4. $\dfrac{18(a^2 - 49)}{3a - 21} = \dfrac{18(a+7)(a-7)}{3(a-7)}$ Factor.

   $= 6(a+7)$ Reduce.

5. $\dfrac{w}{7} \cdot \dfrac{4}{y} \cdot \dfrac{z}{5} = \dfrac{4wz}{35y}$ There are no common factors, so multiply the numerators and multiply the denominators.

6. $\dfrac{7}{y} \div \dfrac{w}{6} = \dfrac{7}{y} \cdot \dfrac{6}{w}$ Multiply by the reciprocal of the divisor.

   $= \dfrac{42}{yw}$ Multiply.

7. $\dfrac{7r}{r-5} + \dfrac{2r}{5-r}$

   $= \dfrac{7r}{r-5} + \dfrac{2r}{-(r-5)}$ $5 - r = -(r-5)$.

Chapter 5 Test

$= \dfrac{7r}{r-5} + \dfrac{-2r}{r-5}$ $\qquad\qquad\qquad\qquad\dfrac{a}{-b} = \dfrac{-a}{b}.$

$= \dfrac{7r + (-2r)}{r-5}$ $\qquad\qquad\qquad\qquad$ Add.

$= \dfrac{5r}{r-5}$

8. $\dfrac{33t^2}{18pq} \cdot \dfrac{45q}{22t}$

$= \dfrac{3t}{2p} \cdot \dfrac{5}{2}$ $\qquad\qquad\qquad\qquad$ Reduce.

$= \dfrac{15t}{4p}$ $\qquad\qquad\qquad\qquad$ Multiply.

9. $\qquad\quad w - 5 = w - 5$ $\qquad\qquad$ Completely factor each polynomial.
$\qquad\quad w^2 - 25 = (w-5)(w+5)$
$w^2 - 10w + 25 = (w-5)^2$

$\qquad\qquad\text{LCM} = (w+5)(w-5)^2$ $\qquad$ Write the product of the largest power of each factor.

10. $\dfrac{1}{2y} + \dfrac{1}{3y} + \dfrac{1}{y}$

$= \dfrac{1}{2y} \cdot \dfrac{3}{3} + \dfrac{1}{3y} \cdot \dfrac{2}{2} + \dfrac{1}{y} \cdot \dfrac{6}{6}$ $\qquad$ Build each fraction to have the common denominator $6y$.

$= \dfrac{3}{6y} + \dfrac{2}{6y} + \dfrac{6}{6y}$

$= \dfrac{3 + 2 + 6}{6y}$ $\qquad\qquad\qquad\qquad$ Add.

$= \dfrac{11}{6y}$

11. $\dfrac{90 \text{ miles}}{1 \text{ hour}}$

$= \dfrac{90 \text{ mi}}{1 \text{ hr}} \cdot \dfrac{5280 \text{ ft}}{1 \text{ mi}} \cdot \dfrac{1 \text{ hr}}{60 \text{ min}} \cdot \dfrac{1 \text{ min}}{60 \text{ sec}}$ $\qquad$ There are 5280 feet in 1 mile.

$= \dfrac{90 \cdot 5280 \cdot 1 \cdot 1 \text{ ft}}{1 \cdot 1 \cdot 60 \cdot 60 \text{ sec}}$ $\qquad\qquad$ Eliminate the common measures.

$= \dfrac{132 \text{ ft}}{1 \text{ sec}}$

12.
$$\frac{4}{y-3} - \frac{3}{3-y} = 1$$

$$\frac{4}{y-3} - \frac{3}{-(y-3)} = 1 \qquad 3 - y = -(y-3).$$

$$\frac{4}{y-3} - \frac{-3}{y-3} = 1 \qquad \frac{a}{-b} = \frac{-a}{b}.$$

$$(y-3)\left(\frac{4}{y-3} - \frac{-3}{y-3}\right) = (y-3)(1) \qquad \text{Multiply both sides by the LCM, } y-3.$$

$$(y-3)\left(\frac{4}{y-3}\right) - (y-3)\left(\frac{-3}{y-3}\right) = (y-3)(1) \qquad \text{Distributive property.}$$

$$4 + 3 = y - 3 \qquad \text{Reduce and multiply.}$$
$$7 = y - 3$$
$$10 = y$$

{10}

13. $\dfrac{x+1}{2x-2} + \dfrac{1}{x-1} - \dfrac{x}{2x+2}$

$= \dfrac{x+1}{2(x-1)} + \dfrac{1}{x-1} - \dfrac{x}{2(x+1)}$ Factor the denominators.

$= \dfrac{x+1}{2(x-1)} \cdot \dfrac{x+1}{x+1} + \dfrac{1}{x-1} \cdot \dfrac{2(x+1)}{2(x+1)} - \dfrac{x}{2(x+1)} \cdot \dfrac{x-1}{x-1}$ Build each fraction to have the common denominator $2(x-1)(x+1)$.

$= \dfrac{x^2+2x+1}{2(x-1)(x+1)} + \dfrac{2x+2}{2(x-1)(x+1)} - \dfrac{x^2-x}{2(x-1)(x+1)}$

$= \dfrac{x^2+2x+1+2x+2-(x^2-x)}{2(x-1)(x+1)}$ Add and subtract.

$= \dfrac{x^2+4x+3-x^2+x}{2(x-1)(x+1)}$

$= \dfrac{5x+3}{2(x-1)(x+1)}$

14. $\dfrac{x^2-9}{x^2-1} \cdot \dfrac{x^2+2x+1}{x^2-2x-3}$

$= \dfrac{(x+3)(x-3)}{(x+1)(x-1)} \cdot \dfrac{(x+1)(x+1)}{(x-3)(x+1)}$ Factor.

$= \dfrac{x+3}{x-1} \cdot \dfrac{1}{1}$ Reduce.

$= \dfrac{x+3}{x-1}$ Multiply.

15. $\dfrac{x^2y^2 - xy}{4x - 4y} \div \dfrac{3xy - 3}{8x - 8y}$

$= \dfrac{x^2y^2 - xy}{4x - 4y} \cdot \dfrac{8x - 8y}{3xy - 3}$ Multiply by the reciprocal of the divisor.

$= \dfrac{xy(xy - 1)}{4(x - y)} \cdot \dfrac{8(x - y)}{3(xy - 1)}$ Factor.

$= \dfrac{xy}{1} \cdot \dfrac{2}{3}$ Reduce.

$= \dfrac{2xy}{3}$ Multiply.

16. $\dfrac{x - \dfrac{9}{x}}{\dfrac{3}{x} + 1}$

$= \dfrac{x - \dfrac{9}{x}}{\dfrac{3}{x} + 1} \cdot \dfrac{x}{x}$ Multiply by the LCM, x.

$= \dfrac{x^2 - 9}{3 + x}$ Simplify.

$= \dfrac{(x + 3)(x - 3)}{3 + x}$ Factor the numerator.

$= x - 3$ Reduce. $x + 3 = 3 + x$.

17. $\dfrac{y}{y + 3} = \dfrac{y + 4}{y + 9}$

$(y + 3)(y + 9)\left(\dfrac{y}{y + 3}\right) = (y + 3)(y + 9)\left(\dfrac{y + 4}{y + 9}\right)$ Multiply both sides by the LCM, $(y + 3)(y + 9)$.

$y^2 + 9y = y^2 + 7y + 12$ Reduce and multiply.
$9y = 7y + 12$
$2y = 12$
$y = 6$

$\{6\}$

18. $\dfrac{3x}{x^2 + x - 2} - \dfrac{2}{x + 2}$

$= \dfrac{3x}{(x + 2)(x - 1)} - \dfrac{2}{x + 2}$ Factor.

$= \dfrac{3x}{(x + 2)(x - 1)} - \dfrac{2}{x + 2} \cdot \dfrac{x - 1}{x - 1}$ Build each fraction to the common denominator, $(x + 2)(x - 1)$.

$= \dfrac{3x}{(x + 2)(x - 1)} - \dfrac{2x - 2}{(x + 2)(x - 1)}$

$$= \frac{3x - (2x - 2)}{(x + 2)(x - 1)}$$ Subtract.

$$= \frac{3x - 2x + 2}{(x + 2)(x - 1)}$$

$$= \frac{x + 2}{(x + 2)(x - 1)}$$

$$= \frac{1}{x - 1}$$ Reduce.

19. $\dfrac{\text{miles}}{\text{inches}}$: $\dfrac{100}{\frac{5}{6}} = \dfrac{x}{4}$ Set up ratios as miles to inches.

$$\frac{5}{5}x = 400$$ Cross multiply.

$$x = 480$$

480 miles

20. $A = kq$ Equation of direct variation.
$210 = k(1)$ Replace A with 210 and q with 1.
$k = 210$ Solve for k.
$A = 210q$ Equation of variation substituting 210 for k.
$1155 = 210q$ Replace A with 1155.
$5.5 = q$ Solve for q.

5.5 qt

21. $f = \dfrac{k}{l}$ Equation of inverse variation.

$42 = \dfrac{k}{12}$ Replace f with 42 and l with 12.

$504 = k$ Solve for k.

$f = \dfrac{504}{l}$ Equation of variation substituting 504 for k.

$f = \dfrac{504}{18}$ Replace l with 18.

$f = 28$ Solve for f.

28 vibrations per sec

EXERCISES 6.1

1. $2x + y = 0$
    $2(3) + (-6) = 0$
    $6 - 6 = 0$
    $0 = 0$

    Replace x with 3 and y with -6.
    True.

    The equation is true; $(3, -6)$ is a solution.

5. $4x - 3y = -36$
    $4(-3) - 3(8) = -36$
    $-12 - 24 = -36$
    $-36 = -36$

    Replace x with -3 and y with 8.
    True.

    The equation is true; $(-3, 8)$ is a solution.

    Three units left and two units up.

9.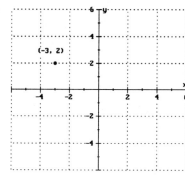

13.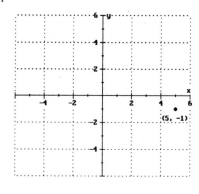

    Five units right and one unit down.

17. $C(-3, 1)$

    Three units left and one unit up.

21. $G(4, -2)$

    Four units right and two units down.

25. $3x - 7y = -72$
    $3(-4) - 7(-12) = -72$
    $-12 + 84 = -72$
    $72 = -72$

    The equation is false; $(-4, -12)$ is not a solution.

29.

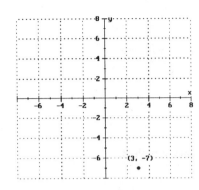

Three units right and seven units down.

33.

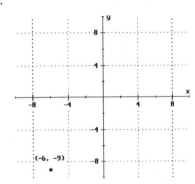

Six units left and nine units down.

37. $A(-4, 3)$

Four units left and three units up.

41. $E(3, 2)$

Three units right and two units up.

45. $\quad\quad 2x - 2y = 5$

$2\left(\dfrac{3}{2}\right) - 2(-1) = 5$

$\quad\quad 3 + 2 = 5$
$\quad\quad\quad\quad 5 = 5$

Replace x with $\dfrac{3}{2}$ and y with -1.

True.

The equation is true; $\left(\dfrac{3}{2}, -1\right)$ is a solution.

49. $\quad\quad 60x - 40y = 111$

$60\left(\dfrac{9}{4}\right) - 40\left(\dfrac{3}{5}\right) = 111$

$\quad\quad 135 - 24 = 111$
$\quad\quad\quad\quad 111 = 111$

Replace x with $\dfrac{9}{4}$ and y with $\dfrac{3}{5}$.

True.

The equation is true; $\left(\dfrac{9}{4}, \dfrac{3}{5}\right)$ is a solution.

Exercises 6.1   159

53.

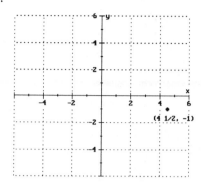

Four and one-half units right and one unit down.

57.

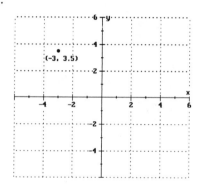

Three units left and three and one-half units up.

61. $C(-1, -2.5)$

One unit left and two and one-half units down.

65. $G(2.5, -2)$

Two and one-half units right and two units down.

69. $(6, D)$

73. $(5, D)$

77. One possible answer: From the origin move four units left and then seven units up and you have located $(-4, 7)$.

85. $25x^2 - 49y^2 = (5x)^2 - (7y)^2$

               $= (5x + 7y)(5x - 7y)$

Write $25x^2$ as the square of $5x$ and $49y^2$ as the square of $7y$.
To factor, write the sum and difference of the terms that were squared.

89. $\dfrac{12mn}{5n^2} \cdot \dfrac{30m^2}{4mn} = \dfrac{3m}{n} \cdot \dfrac{6m}{n}$

                   $= \dfrac{18m^2}{n^2}$

Reduce.

Multiply.

EXERCISES 6.2

1. $x + y = 12$
   $6 + y = 12$
   $y = 6$ Replace x with 6.

   (6, 6) is a solution.

5. $x + y = 12$
   $-7 + y = 12$
   $y = 19$ Replace x with -7.

   (-7, 19) is a solution.

9. $x - y = 15$
   $x - 6 = 15$
   $x = 21$ Replace y with 6.

   (21, 6) is a solution.

13. $x - y = 15$
    $x - 15 = 15$
    $x = 30$ Replace y with 15.

    (30, 15) is a solution.

17. $y = \dfrac{1}{2}x$

x	y
-6	-3
0	0
6	3

    $y = \dfrac{1}{2}(-6) = -3.$

    $y = \dfrac{1}{2}(0) = 0.$

    $y = \dfrac{1}{2}(6) = 3.$

    Plot the points and connect with a straight edge.

    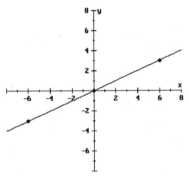

21. $x - y = 4$

x	y
0	-4
4	0
2	-2

    $0 - y = 4;\ y = -4.$
    $x - 0 = 4;\ x = 4.$
    $2 - y = 4;\ y = -2.$

Exercises 6.2   161

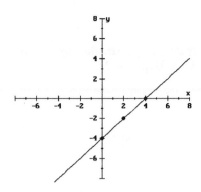

Plot the points and connect with a straight edge.

25. $2x - y = 6$
     $2(0) - y = 6$ Replace x with 0.
         $-y = 6$
          $y = -6$

     $(0, -6)$ is a solution.

29. $2x - y = 6$
     $2(6.5) - y = 6$ Replace x with 6.5.
        $13 - y = 6$
            $-y = -7$
             $y = 7$

     $(6.5, 7)$ is a solution.

33. $3x - y = 8$
     $3x - 0 = 8$ Replace y with 0.
          $3x = 8$
           $x = \dfrac{8}{3}$

     $\left(\dfrac{8}{3}, 0\right)$ is a solution.

37. $3x - y = 8$

     $3x - \dfrac{5}{2} = 8$ Replace y with $\dfrac{5}{2}$.

     $6x - 5 = 16$ Multiply both sides by 2 to eliminate the fractions.
          $6x = 21$
           $x = \dfrac{7}{2}$

     $\left(\dfrac{7}{2}, \dfrac{5}{2}\right)$ is a solution.

41. $x + 2y = 6$

x	y
0	3
6	0
2	2

Find one point by letting $x = 0$, another by letting $y = 0$, and a third by letting $x = 2$.

Plot the points and connect with a straight edge.

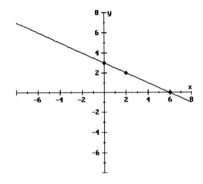

45. $3x + 4y = 12$

x	y
0	3
4	0
-4	6

Find one point by letting $x = 0$, another by letting $y = 0$, and a third by letting $x = -4$.

Plot the points and connect with a straight edge.

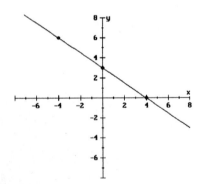

49. $3x + 5y = -15$

x	y
0	-3
-5	0
5	-6

Find one point by letting $x = 0$, another by letting $y = 0$, and a third by letting $x = 5$.

Exercises 6.2

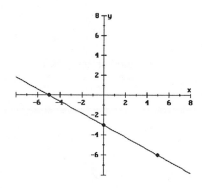

Plot the points and connect with a straight edge.

53. $x - 4y = 8$

x	y
0	-2
8	0
4	-1

Find one point by letting $x = 0$, another by letting $y = 0$, and a third by letting $x = 4$.

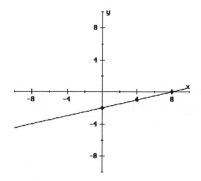

Plot the points and connect with a straight edge.

57. $4x - 5y = 18$

x	y
0	-3.6
4.5	0
7	2

Find one point by letting $x = 0$, another by letting $y = 0$ and a third by letting $y = 2$.

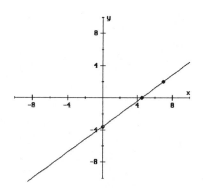

Plot the points and connect with a straight edge.

61. $5x - 4y = 20$

x	y
0	-5
4	0
-4	-10

Find one point by letting $x = 0$, another by letting $y = 0$ and a third by letting $x = -4$.

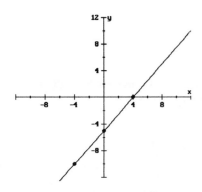

Plot the points and connect with a straight edge.

65. $0.4x + 0.3y = 1.2$
$4x + 3y = 12$

Multiply both sides by 10.

x	y
0	4
3	0
6	-4

Find one point by letting $x = 0$, another by letting $y = 0$ and a third by letting $x = 6$.

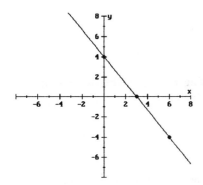

Plot the points and connect with a straight edge.

69. 1st number: x
2nd number: y

$x = 2y - 2$
$47 = 2y - 2$
$49 = 2y$
$24.5 = y$

Replace x with 47.

The second number is 24.5.

Exercises 6.2

73. units of Munch: x
units of Good AM: y

Grams per unit · # of units + Grams per unit · # of units = Total grams

$$20x + 30y = 180$$
$$20x + 30(2) = 180 \quad \text{Replace } y \text{ with 2.}$$
$$20x + 60 = 180$$
$$20x = 120$$
$$x = 6$$

6 units of Munch

77. $y = 45000 - 3600x$

x	y
0	45000
5	27000
12	1800

Find one point by letting $x = 0$, another by letting $x = 5$ and a third by letting $x = 12$.

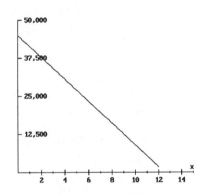

Plot the points and connect with a straight edge.

81. One possible answer: An equation in two variables has an infinite number of solutions because you can choose any real number for one of the variables and solve for the other variable.

85. $2y - 12 = 0$
$2y = 12$
$y = 6$

x	y
0	6
1	6
2	6

No matter what value is chosen for x, the corresponding y-value is 6.

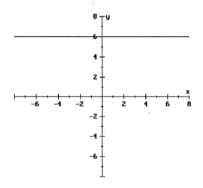

89. $x^2 + 4x - 32 = 0$
    $(x + 8)(x - 4) = 0$ Factor the left side.
    $x + 8 = 0$ or $x - 4 = 0$ Zero-product property.
    $x = -8$ or $x = 4$

    $\{-8, 4\}$

93. $\dfrac{5x^2}{6y^2} \cdot \dfrac{8y}{15x^3} = \dfrac{1}{3y} \cdot \dfrac{4}{3x}$ Reduce.

    $= \dfrac{4}{9xy}$ Multiply.

97. $\dfrac{x^2 + x - 30}{2x^2 - 15x + 7} \cdot \dfrac{2x^2 + 3x - 2}{3x^2 - 13x - 10} \cdot \dfrac{3x^2 - 19x - 14}{x^2 - 4x - 12}$

    $= \dfrac{(x + 6)(x - 5)}{(2x - 1)(x - 7)} \cdot \dfrac{(2x - 1)(x + 2)}{(3x + 2)(x - 5)} \cdot \dfrac{(3x + 2)(x - 7)}{(x - 6)(x + 2)}$ Factor.

    $= \dfrac{x + 6}{1} \cdot \dfrac{1}{1} \cdot \dfrac{1}{x - 6}$ Reduce.

    $= \dfrac{x + 6}{x - 6}$ Multiply.

EXERCISES 6.3

1. $4x + y = 11$

    y-intercept:
    $4(0) + y = 11$ Replace x with 0. All points on the y-axis have an x-value of 0.
    $y = 11$ $(0, 11)$.

    x-intercept:
    $4x + 0 = 11$ Replace y with 0. All points on the x-axis have a y-value of 0.
    $4x = 11$
    $x = \dfrac{11}{4}$ $\left(\dfrac{11}{4}, 0\right)$.

Exercises 6.3

5. $2x + 9y = 18$

    y-intercept:
    $2(0) + 9y = 18$
    $9y = 18$
    $y = 2$

    Replace x with 0. All points on the y-axis have an x-value of 0.
    $(0, 2)$.

    x-intercept:
    $2x + 9(0) = 18$
    $2x = 18$
    $x = 9$

    Replace y with 0. All points on the x-axis have a y-value of 0.
    $(9, 0)$.

9. $m = \dfrac{y_2 - y_1}{x_2 - x_1}$

    Formula for slope.

    $m = \dfrac{2 - (-1)}{-10 - 6}$

    Replace y_2 with 2, y_1 with -1, x_2 with -10 and x_1 with 6.

    $m = -\dfrac{3}{16}$

    Simplify.

13. $y = -\dfrac{1}{3}x + 7$

    $y = mx + b$.

    $m = -\dfrac{1}{3}$

    $(0, 7)$

    $b = 7$.

17. $4y = -7x - 8$

    $y = -\dfrac{7}{4}x - 2$

    Solve for y.

    $m = -\dfrac{7}{4}$

    $y = mx + b$.

    $(0, -2)$

    $b = -2$.

21. $5x - 7y = 15$

    y-intercept:
    $5(0) - 7y = 15$
    $-7y = 15$

    Replace x with 0.

    $y = -\dfrac{15}{7}$

    $\left(0, -\dfrac{15}{7}\right)$

    x-intercept:
    $5x - 7(0) = 15$
    $5x = 15$
    $x = 3$

    Replace y with 0.
    $(3, 0)$.

25. $m = \dfrac{y_2 - y_1}{x_2 - x_1}$ Formula for slope.

$m = \dfrac{-4 - (-1)}{5 - (-2)}$ Replace y_2 with -4, y_1 with -1, x_2 with 5 and x_1 with -2.

$= -\dfrac{3}{7}$ Simplify.

29. $m = \dfrac{y_2 - y_1}{x_2 - x_1}$ Formula for slope.

$m = \dfrac{-2 - 3}{-12 - 0}$ Replace y_2 with -2, y_1 with 3, x_2 with -12 and x_1 with 0.

$= \dfrac{5}{12}$ Simplify.

33. $-3x - 4y = 9$

$-4y = 3x + 9$ Solve for y.

$= -\dfrac{3}{4}x - \dfrac{9}{4}$ $y = mx + b$.

$m = -\dfrac{3}{4}$

$\left(0, -\dfrac{9}{4}\right)$ $b = -\dfrac{9}{4}$.

37. $\dfrac{1}{3}x + \dfrac{1}{2}y = -5$

y-intercept:

$\dfrac{1}{3}(0) + \dfrac{1}{2}y = -5$ Replace x with 0.

$\dfrac{1}{2}y = -5$

$y = -10$ $(0, -10)$.

x-intercept:

$\dfrac{1}{3}x + \dfrac{1}{2}(0) = -5$ Replace y with 0.

$\dfrac{1}{3}x = -5$

$x = -15$ $(-15, 0)$.

Exercises 6.3

41. $0.5x - 2.5y = 1$

    y-intercept:
$$0.5(0) - 2.5y = 1$$
$$-2.5y = 1$$
$$y = -0.4$$

Replace x with 0.

$(0, -0.4)$.

    x-intercept:
$$0.5x - 2.5(0) = 1$$
$$0.5x = 1$$
$$x = 2$$

Replace y with 0.

$(2, 0)$.

45. $m = \dfrac{y_2 - y_1}{x_2 - x_1}$

Formula for slope.

$m = \dfrac{-\dfrac{1}{2} - \dfrac{3}{4}}{\dfrac{2}{3} - \dfrac{1}{6}}$

Replace y_2 with $-\dfrac{1}{2}$, y_1 with $\dfrac{3}{4}$, x_2 with $\dfrac{2}{3}$ and x_1 with $\dfrac{1}{6}$.

$m = \dfrac{-\dfrac{5}{4}}{\dfrac{1}{2}}$

Simplify.

$m = -\dfrac{5}{2}$

49. $100y = 7x$

$y = \dfrac{7}{100}x$

Solve for y.

$m = \dfrac{7}{100}$

$y = mx + b$.

53. $m = \dfrac{\text{rise}}{\text{run}}$

$m = \dfrac{3}{4}$

Replace rise with 3 and run with 4.

57. One possible answer: The slope is negative since the line falls from left to right.

61. $y - 6 = 0$
$$y = 6$$
$$y = 0x + 6$$
$$m = 0$$
$$(0, 6)$$

Solve for y.
$y = mx + b$.
$b = 6$.

65. $\dfrac{14x^3}{25xy} \cdot \dfrac{15y^3}{7x^2y}$

$= \dfrac{2}{5} \cdot \dfrac{3y}{1}$ Reduce.

$= \dfrac{6y}{5}$ Multiply.

69. $\dfrac{7b^2 - 14b}{b^2 - 4b} \cdot \dfrac{b^3 - 16b}{9b^2 - 18b}$

$= \dfrac{7b(b - 2)}{b(b - 4)} \cdot \dfrac{b(b + 4)(b - 4)}{9b(b - 2)}$ Factor.

$= \dfrac{7}{1} \cdot \dfrac{b + 4}{9}$ Reduce.

$= \dfrac{7(b + 4)}{9}$ Multiply.

73. width: x
length: $x + 5$

$\quad\quad x(x + 5) = 204$ $wl = A$.
$\quad\quad x^2 + 5x = 204$ Multiply.
$\quad x^2 + 5x - 204 = 0$ Subtract 204 from both sides.
$(x - 12)(x + 17) = 0$ Factor.
$x - 12 = 0 \quad \text{or} \quad x + 17 = 0$ Zero-product property.
$\quad x = 12 \quad \text{or} \quad \quad x = -17$

Since $x > 0$, $x = 12$
and $x + 5 = 12 + 5 = 17$.
12 in. × 17 in.

EXERCISES 6.4

1.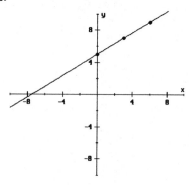

Locate (0, 5). Go up 2 units and to the right 3 units. Repeat this step from the new point (3, 7) to get a third point. Draw the line through the three points.

Exercises 6.4

5.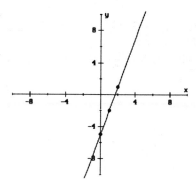

Locate (0, −5). Go up 3 units and to the right 1 unit. Repeat this step from the new point (1, −2) to get a third point. Draw the line through the three points.

9.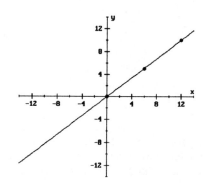

Locate (0, 0). Go up 5 units and to the right 6 units. Repeat this step from the new point (6, 5) to get a third point. Draw the line through the three points.

13.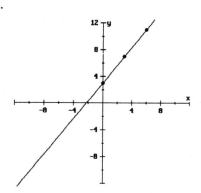

Locate (0, 3). Go up 4 units and to the right 3 units. Repeat this step from the new point (3, 7) to get a third point. Draw the line through the three points.

17. $4x + 5y = 20$
$5y = 4x + 20$
$y = -\dfrac{4}{5}x + 4$
$m = -\dfrac{4}{5}$; (0, 4)

Solve for y.

17.

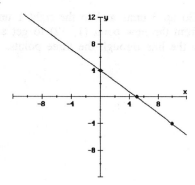

Locate (0, 4). Go down 4 units and to the right 5 units. Repeat this step from the new point (5, 0) to get a third point. Draw the line through the three points.

21. $2x - 3y = 6$
$-3y = -2x + 6$

$y = \dfrac{2}{3}x - 2$

$m = \dfrac{2}{3};\ (0, -2)$

Solve for y.

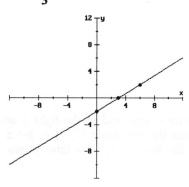

Locate (0, −2). Go up 2 units and to the right 3 units. Repeat this step from the new point (3, 0) to get a third point. Draw the line through the three points.

25. $5x + 4y = -12$
$4y = -5x - 12$

$y = -\dfrac{5}{4}x - 3$

$m = -\dfrac{5}{4};\ (0, -3)$

Solve for y.

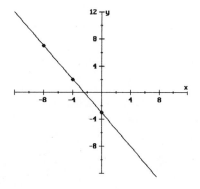

Locate (0, −3). Go up 5 units and to the left 4 units. Repeat this step from the new point (−4, 2) to get a third point. Draw the line through the three points.

Exercises 6.4

29. $2x + y = 4$
    $y = -2x + 4$ Solve for y.

    $m = -2;\ (0, 4)$

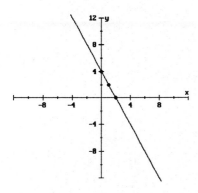

Locate $(0, 4)$. Go down 2 units and to the right 1 unit. Repeat this step from the new point $(1, 2)$ to get a third point. Draw the line through the three points.

33. $x + \dfrac{2}{3}y = -1$

    $\dfrac{2}{3}y = -x - 1$ Solve for y.

    $y = -\dfrac{3}{2}x - \dfrac{3}{2}$

    $m = -\dfrac{3}{2};\ \left(0, -\dfrac{3}{2}\right)$

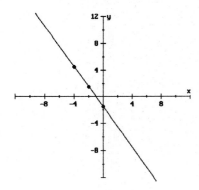

Locate $\left(0, -\dfrac{3}{2}\right)$. Go up 3 units and to the left 2 units. Repeat this step from the new point $\left(-2, \dfrac{3}{2}\right)$ to get a third point. Draw the line through the three points.

37. $-x + \dfrac{1}{2}y = 1$

    $\dfrac{1}{2}y = x + 1$ Solve for y.

    $y = 2x + 2$

    $m = 2;\ (0, 2)$

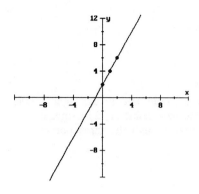

Locate (0, 2). Go up 2 units and to the right 1 unit. Repeat this step from the new point (1, 4) to get a third point. Draw the line through the three points.

41. $-\frac{1}{3}x - y = -2$

$-y = \frac{1}{3}x - 2$ Solve for y.

$y = -\frac{1}{3}x + 2$

$m = -\frac{1}{3}$; (0, 2)

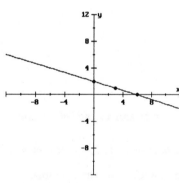

Locate (0, 2). Go down 1 unit and to the right 3 units. Repeat this step from the new point (3, 1) to get a third point. Draw the line through the three points.

45. $\frac{1}{2}x - 2y = \frac{4}{5}$

$-2y = -\frac{1}{2}x + \frac{4}{5}$ Solve for y.

$y = \frac{1}{4}x - \frac{2}{5}$

$m = \frac{1}{4}$; $\left(0, -\frac{2}{5}\right)$

Exercises 6.4

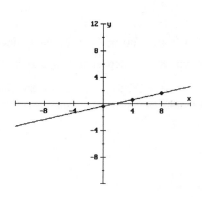

Locate $\left(0, -\frac{2}{5}\right)$. Go down 1 unit and to the right 4 units. Repeat this step from the new point $\left(4, \frac{3}{5}\right)$ to get a third point. Draw the line through the three points.

49. $y = 10000 + 5x$

   $m = 5;\ (0, 10{,}000)$

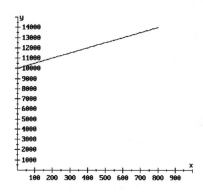

Let each unit on the y-axis equal 1000 and each unit on the x-axis equal 100.

53. $y = 25000 + 125x$

   $m = 125;\ (0, 25000)$

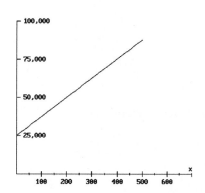

Total cost = Fixed cost + Variable cost.

Let each unit on the y-axis equal 25,000 and each unit on the x-axis equal 100.

57. (0, −2); $m = 0$

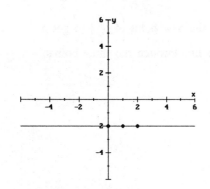

Locate (0, −2). Go 0 units up and 1 unit right, since $m = 0$ can be written as $\frac{0}{1}$. Repeat this step from the new point (1, −2) to get a third point. Draw the line through the three points.

61. $y = \frac{3}{5}x + 5$

$y = \frac{3}{5}(15) + 5$ Replace x with 15.

$y = 9 + 5$ Solve for y.
$y = 14$

65. $6x - 11y = 22$
   $6(22) - 11y = 22$ Replace x with 22.
   $132 - 11y = 22$ Solve for y.
        $-11y = -110$
           $y = 10$

69. $2w + 2l = P$ Formula for perimeter.
   $2w + 2(32) = 84$ Replace l with 32 and P with 84.
   $2w + 64 = 84$ Solve for w.
       $2w = 20$
        $w = 10$

10 ft

EXERCISES 6.5

1. $y - y_1 = m(x - x_1)$ Point-slope form.
   $y - 0 = 4(x - 3)$ Substitute 0 for y_1, 3 for x_1 and 4 for m.

     $y = 4x - 12$ Simplify.
     $12 = 4x - y$ Standard form.

5. $y - y_1 = m(x - x_1)$ Point-slope form.
   $y - 1 = -2[x - (-4)]$ Substitute 1 for y_1, −4 for x_1 and −2 for m.

   $y - 1 = -2(x + 4)$ Simplify.
   $y - 1 = -2x - 8$
   $2x + y = -7$ Standard form.

Exercises 6.5

9. $m = \dfrac{y_2 - y_1}{x_2 - x_1}$ Formula for the slope.

       $m = \dfrac{5 - 3}{2 - 1}$ Substitute: (x_1, y_1) is $(1, 3)$ and (x_2, y_2) is $(2, 5)$.

       $m = 2$ Slope of the line.

       $y - y_1 = m(x - x_1)$ Point-slope form.
       $y - 3 = 2(x - 1)$ Substitute: (x_1, y_1) is $(1, 3)$ and m is 2.

       $y - 3 = 2x - 2$ Simplify.
       $-1 = 2x - y$ Standard form.

13. $y - y_1 = m(x - x_1)$ Point-slope form.

       $y - 6 = \dfrac{1}{2}(x - 2)$ Substitute 6 for y_1, 2 for x_1 and $\dfrac{1}{2}$ for m.

       $2y - 12 = x - 2$ Multiply both sides by 2.
       $-10 = x - 2y$ Standard form.

17. $y - y_1 = m(x - x_1)$ Point-slope form.

       $y - 1 = -\dfrac{2}{5}(x - 1)$ Substitute 1 for y_1, 1 for x_1 and $-\dfrac{2}{5}$ for m.

       $5y - 5 = -2(x - 1)$ Multiply both sides by 5.
       $5y - 5 = -2x + 2$ Simplify.
       $2x + 5y = 7$ Standard form.

21. $m = \dfrac{y_2 - y_1}{x_2 - x_1}$ Formula for the slope.

       $m = \dfrac{-9 - 4}{5 - (-3)}$ Substitute: (x_1, y_1) is $(-3, 4)$ and (x_2, y_2) is $(5, -9)$.

       $m = -\dfrac{13}{8}$ Simplify.

       $y - y_1 = m(x - x_1)$ Point-slope form.

       $y - 4 = -\dfrac{13}{8}[x - (-3)]$ Substitute: (x_1, y_1) is $(-3, 4)$ and m is $-\dfrac{13}{8}$.

       $8y - 32 = -13(x + 3)$ Multiply both sides by 8.
       $8y - 32 = -13x - 39$ Simplify.
       $13x + 8y = -7$ Standard form.

25. $y - y_1 = m(x - x_1)$ Point-slope form.

       $y - \dfrac{5}{13} = -\dfrac{3}{13}(x - 4)$ Substitute: (x_1, y_1) is $\left(4, \dfrac{5}{13}\right)$ and m is $-\dfrac{3}{13}$.

       $13y - 5 = -3(x - 4)$ Multiply both sides by 13.
       $13y - 5 = -3x + 12$ Simplify.
       $3x + 13y = 17$ Standard form.

29. $y - y_1 = m(x - x_1)$ Point-slope form.

$y - \left(-\dfrac{3}{2}\right) = -\dfrac{9}{5}(x - 4)$ Substitute: (x_1, y_1) is $\left(4, -\dfrac{3}{2}\right)$ and m is $-\dfrac{9}{5}$.

$10y + 15 = -18(x - 4)$ Multiply both sides by 10.
$10y + 15 = -18x + 72$ Simplify.
$18x + 10y = 57$ Standard form.

33. $m = \dfrac{y_2 - y_1}{x_2 - x_1}$ Formula for the slope.

$m = \dfrac{-\dfrac{1}{3} - \dfrac{2}{3}}{-\dfrac{2}{9} - \dfrac{5}{9}}$ Substitute: (x_1, y_1) is $\left(\dfrac{5}{9}, \dfrac{2}{3}\right)$ and (x_2, y_2) is $\left(-\dfrac{2}{9}, -\dfrac{1}{3}\right)$.

$m = \dfrac{-1}{-\dfrac{7}{9}}$ Simplify.

$m = \dfrac{9}{7}$ Slope of the line.

$y - y_1 = m(x - x_1)$ Point-slope form.

$y - \dfrac{2}{3} = \dfrac{9}{7}\left(x - \dfrac{5}{9}\right)$ Substitute: (x_1, y_1) is $\left(\dfrac{5}{9}, \dfrac{2}{3}\right)$ and m is $\dfrac{9}{7}$.

$21y - 14 = 27\left(x - \dfrac{5}{9}\right)$ Multiply both sides by 21.

$21y - 14 = 27x - 15$ Simplify.
$1 = 27x - 21y$ Standard form.

37. (1, 50); (2, 60)

$m = \dfrac{y_2 - y_1}{x_2 - x_1}$ Formula for the slope.

$m = \dfrac{60 - 50}{2 - 1}$ Substitute: (x_1, y_1) is (1, 50) and (x_2, y_2) is (2, 60).

$m = 10$ Simplify.

$y - y_1 = m(x - x_1)$ Point-slope form.
$y - 50 = 10(x - 1)$ Substitute: (x_1, y_1) is (1, 50) and m is 10.

$y - 50 = 10x - 10$ Simplify.
$-40 = 10x - y$ Standard form.

$-40 = 10(7) - y$ Replace x with 7.
$-40 = 70 - y$ Solve for y.
$-110 = -y$
$110 = y$

110 enrollments.

Exercises 6.5

41. (2, 40); (4, 10)

$m = \dfrac{y_2 - y_1}{x_2 - x_1}$ Formula for the slope.

$m = \dfrac{40 - 10}{2 - 4}$ Substitute: (x_1, y_1) is (4, 10) and (x_2, y_2) is (2, 40).

$m = \dfrac{30}{-2}$ Simplify.

$m = -15$ Slope of the line

$y - y_1 = m(x - x_1)$ Point-slope form.
$y - 10 = -15(x - 4)$ Substitute: (x_1, y_1) is (4, 10) and m is -15.
$y - 10 = -15x + 60$ Simplify.
$15x + y = 70$ Standard form.

$15x + 0 = 70$ Replace y with 0.

$x = 4\dfrac{2}{3}$ Solve for x.

$4\dfrac{2}{3}$ sec

45. $m = \dfrac{y_2 - y_1}{x_2 - x_1}$ Formula for the slope.

$\dfrac{2}{3} = \dfrac{5 - 0}{-4 - k}$ Substitute: (x_1, y_1) is $(k, 0)$, (x_2, y_2) is $(-4, 5)$ and m is $\dfrac{2}{3}$.

$2(-4 - k) = 15$ Cross multiply.

$-8 - 2k = 15$ Solve for k.
$-2k = 23$

$k = -\dfrac{23}{2}$

49. $\dfrac{3x}{x - 2} - \dfrac{6}{x - 2} = \dfrac{3x - 6}{x - 2}$ Find the difference of the numerators.

$= \dfrac{3(x - 2)}{x - 2}$ Factor the numerator.

$= 3$ Reduce.

53. $\dfrac{x - 6}{x^2 - 3x - 4} + \dfrac{x + 4}{x - 4} - \dfrac{3}{x + 1}$

$= \dfrac{x - 6}{(x - 4)(x + 1)} + \dfrac{x + 4}{x - 4} - \dfrac{3}{x + 1}$ Factor.

$= \dfrac{x - 6}{(x - 4)(x + 1)} + \dfrac{x + 4}{x - 4} \cdot \dfrac{x + 1}{x + 1} - \dfrac{3}{x + 1} \cdot \dfrac{x - 4}{x - 4}$ Build each fraction to the common denominator $(x + 1)(x - 4)$.

$$= \frac{x-6}{(x-4)(x+1)} + \frac{x^2+5x+4}{(x-4)(x+1)} - \frac{3x-12}{(x-4)(x+1)} \qquad \text{Add and subtract.}$$

$$= \frac{x-6+x^2+5x+4-(3x-12)}{(x-4)(x+1)}$$

$$= \frac{x^2+6x-2-3x+12}{(x-4)(x+1)} \qquad \text{Simplify.}$$

$$= \frac{x^2+3x+10}{(x-4)(x+1)}$$

57. $\dfrac{\dfrac{1}{6}-\dfrac{1}{b}}{\dfrac{1}{4}-\dfrac{1}{b}}$

$$= \frac{\dfrac{1}{6}-\dfrac{1}{b}}{\dfrac{1}{4}-\dfrac{1}{b}} \cdot \frac{12b}{12b} \qquad \text{Multiply the numerator and the denominator by the LCM, } 12b.$$

$$= \frac{2b-12}{3b-12} \qquad \text{Simplify.}$$

$$= \frac{2(b-6)}{3(b-4)} \qquad \text{Factor.}$$

EXERCISES 6.6

1. $x - y < -4$
   $0 - 0 < -4$
   $0 < -4$

   The boundary is the dashed line $x - y = -4$.
   (0, 0) makes a good check point.
   False. Shade the other half-plane.

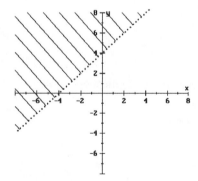

5. $3x - y \le 0$
   $3(1) - 1 \le 0$
   $2 \le 0$

   The boundary is the solid line $3x - y = 0$.
   (1, 1) makes a good check point.
   False. Shade the other half-plane.

Exercises 6.6

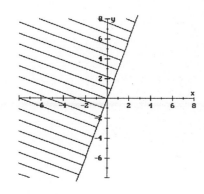

9. $2x - y > 5$
    $2(0) - 0 > 5$
    $0 > 5$

The boundary is the dashed line $2x - y = 5$.
(0, 0) makes a good check point.
False. Shade the other half-plane.

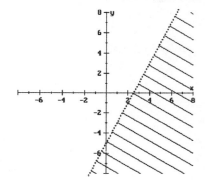

13. $x + y < 3$
    $0 + 0 < 3$
    $0 < 3$

The boundary is the dashed line $x + y = 3$.
(0, 0) makes a good check point.
True. Shade the half-plane containing the origin.

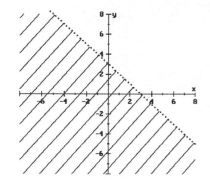

17. $3x - y \le 5$
    $3(0) - 0 \le 5$
    $0 \le 5$

The boundary is the solid line $3x - y = 5$.
(0, 0) is a good check point.
True. Shade the half-plane containing the origin.

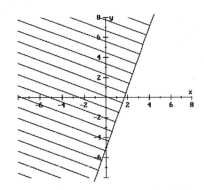

21. $x - 2y \leq -1$
     $0 - 2(0) \leq -1$
     $0 \leq -1$

The boundary is the solid line $x - 2y = -1$.
(0, 0) is a good check point.
False. Shade the other half-plane.

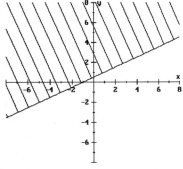

25. $5x + 2y < 2$
     $5(0) + 2(0) < 2$
     $0 < 2$

The boundary is the dashed line $5x + 2y = 2$.
(0, 0) is a good check point.
True. Shade the half-plane containing the origin.

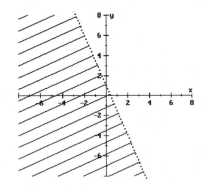

29. $\frac{1}{2}x + \frac{1}{4}y \geq 2$

     $4\left(\frac{1}{2}x + \frac{1}{4}y\right) \geq 4(2)$

     $2x + y \geq 8$

     $2(0) + 0 \geq 8$
     $0 \geq 8$

Clear the fractions.

The boundary is the solid line $2x + y = 8$.

(0, 0) is a good check point.
False. Shade the other half-plane.

Chapter 6 Concept Review

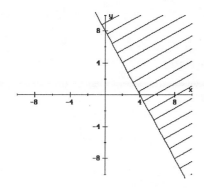

33. $0.3x - 0.4y \geq 2$
     $10(0.3x - 0.4y) \geq 10(2)$
     $3x - 4y \geq 20$
     $3(0) - 4(0) \geq 20$
     $0 \geq 20$

Clear the decimals.
The boundary is the solid line $3x - 4y = 20$.
$(0, 0)$ is a good check point.
False. Shade the other half-plane.

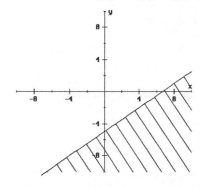

37. One possible answer: A dashed line is used when the inequality symbol is either $<$ or $>$. The points on this boundary line are not included since they would cause an equality.

45. $-4a(b - 3c + 3) + b(2a - 3c + 7) - 5c(-a - 4b + 12)$
    $= -4ab + 12ac - 12a + 2ab - 3bc + 7b + 5ac + 20bc - 60c$
    $= -2ab + 17ac - 12a + 17bc + 7b - 60c$

Multiply.
Combine like terms.

49. $(x + 2)(3x - 5) + (2x - 5)(2x + 5) - (2x - 7)(3x + 4)$
    $= 3x^2 + x - 10 + 4x^2 - 25 - (6x^2 - 13x - 28)$
    $= 3x^2 + x - 10 + 4x^2 - 25 - 6x^2 + 13x + 28$
    $= x^2 + 14x - 7$

Multiply using FOIL.
Change the subtraction to addition.
Combine like terms.

CHAPTER 6 CONCEPT REVIEW

1. False;   It has an unlimited number of solutions.

2. True

3. True

4. False;   It is called the origin.

5. True;

6. True

7. False;   It is not a solution since,

$$2(-1) + 3\left(\frac{3}{2}\right) = 0$$

$$-2 + \frac{9}{2} = 0$$

$$\frac{5}{2} = 0$$

is a false statement.

8. False;   It is located 5 units below the x-axis.

9. True

10. True

11. False;   Most vertical lines do not have a y-intercept.

12. True

13. True

14. False;   The x-value is 0.

15. False;   The slope is undefined.

16. True

17. True

18. True

19. True

20. False;   The equation is $12x + 5y = 11$.

CHAPTER 6 TEST

1. $A(5, 5)$ Right 5, up 5.
      $B(2, -4)$ Right 2, down 4.
      $C(-5, 3)$ Left 5, up 3.
      $D(0, -2)$ Down 2 on y-axis.
      $E(-6, -4)$ Left 6, down 4.

Chapter 6 Test

2.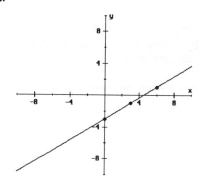
Locate (0, −3). Go up 2 units and to the right 3 units. Repeat this step from the new point (3, −1) to find a third point. Draw a line through the three points.

3. $4x - 5y + 20 = 0$
    $4x - 5(-8) + 20 = 0$ Replace y with -8.
    $4x + 40 + 20 = 0$ Solve for x.
    $4x + 60 = 0$
    $4x = -60$
    $x = -15$

4. $4x - 5y + 20 = 0$
    $4(-10) - 5y + 20 = 0$ Replace x with -10.
    $-40 - 5y + 20 = 0$ Solve for y.
    $-5y - 20 = 0$
    $-5y = 20$
    $y = -4$

5. $2x - 5y = 15$

   y-intercept:
   $2(0) - 5y = 15$ Replace x with 0.
   $-5y = 15$ Solve for y.
   $y = -3$ $(0, -3)$.

   x-intercept:
   $2x - 5(0) = 15$ Replace y with 0.
   $2x = 15$ Solve for x.
   $x = 7.5$ $(7.5, 0)$.

6. $2x - y > -2$
    $2(0) - 0 > -2$ The boundary is the dashed line $2x - y = -2$.
    $0 > -2$ $(0, 0)$ is a good check point.
                                    True. Shade the half-plane containing the origin.

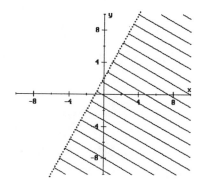

7. $x + 2y = 6$

x	y
0	3
6	0
2	2

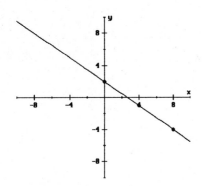

Find one point by letting $x = 0$, another by letting $y = 0$ and a third by letting $x = 2$. Connect the three points with a straight edge.

8. $y = -\frac{3}{4}x + 2$

$m = -\frac{3}{4};\ (0, 2)$

Locate $(0, 2)$. Go down 3 units and to the right 4 units. Repeat this step from the new point $(4, -1)$ to find a third point. Draw a line through the three points.

9. $m = \dfrac{y_2 - y_1}{x_2 - x_1}$ Slope formula.

$m = \dfrac{-8 - (-5)}{3 - (-2)}$ Substitute: (x_1, y_1) is $(-2, -5)$ and (x_2, y_2) is $(3, -8)$.

$m = -\dfrac{3}{5}$ Simplify.

10. $3x = 7y - 6$

$3x + 6 = 7y$ Solve for y.

$\dfrac{3}{7}x + \dfrac{6}{7} = y$ $mx + b = y$.

$m = \dfrac{3}{7}$

Chapter 6 Test

11.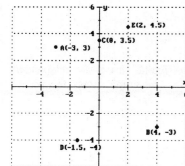

A(−3, 3): Left 3, up 3.
B(4, −3): Right 4, down 3.
C(0, 3.5): Up 3.5 on y-axis.
D(−1.5, −4): Left 1.5, down 4.
E(2, 4.5): Right 2, up 4.5.

12. $3x - 6y \geq -4$
$3(0) - 6(0) \geq -4$
$0 \geq -4$

The boundary is the solid line $3x - 6y = -4$.
(0, 0) is a good check point.
True. Shade the half-plane containing the origin.

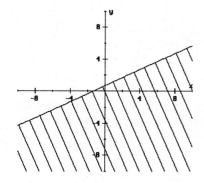

13. $y - y_1 = m(x - x_1)$ Point-slope form.

$y - (-6) = -\dfrac{3}{5}[x - (-1)]$ Substitute: (x_1, y_1) is $(-1, -6)$ and m is $-\dfrac{3}{5}$.

$5y + 30 = -3(x + 1)$ Multiply both sides by 5.
$5y + 30 = -3x - 3$ Simplify.
$3x + 5y = -33$ Standard form.

14. $7y = -2x - 4$
$7(2) = -2(-9) - 4$ Replace x with -9 and y with 2.
$14 = 14$ True.

(−9, 2) is a solution.

15. $m = \dfrac{y_2 - y_1}{x_2 - x_1}$ Formula for the slope.

$m = \dfrac{-6 - 8}{3 - 5}$ Substitute: (x_1, y_1) is $(5, 8)$ and (x_2, y_2) is $(3, -6)$.

$m = 7$ Simplify.

$y - y_1 = m(x - x_1)$ Point-slope form.
$y - 8 = 7(x - 5)$ Substitute: (x_1, y_1) is $(5, 8)$ and m is 7.

$y - 8 = 7x - 35$ Simplify.
$27 = 7x - y$ Standard form.

CHAPTER 7

EXERCISES 7.1

1. l_1: $x + y = 5$
   $y = -x + 5$

   l_2: $x - y = 1$
   $x - 1 = y$

   Since the slopes are different, this is an independent and consistent system. The graph shows the point of intersection.

   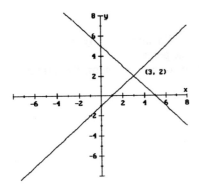

5. l_1: $x + y = -2$
   $y = -x - 2$

   l_2: $x - y = -6$
   $x + 6 = y$

   Since the slopes are different, this is an independent and consistent system. The graph shows the point of intersection.

   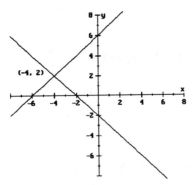

9. l_1: $x + y = 5$
   $y = -x + 5$

   l_2: $-x - y = -6$
   $-x + 6 = y$

   The slope of each line is -1. The y-intercept of l_1 is $(0, 5)$ and the y-intercept of l_2 is $(0, 6)$. Since the slopes are equal and the y-intercepts are different, the lines are parallel. The system has no solution.

   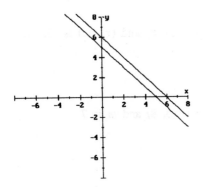

Exercises 7.1

13. l_1: $x + y = 8$
$y = -x + 8$

l_2: $2x + y = 12$
$y = -2x + 12$

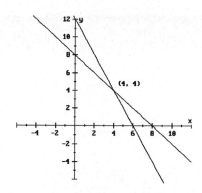

Since the slopes are different, this is an independent and consistent system. The graph shows the point of intersection.

17. l_1: $y = x + 11$
l_2: $y = -x + 3$

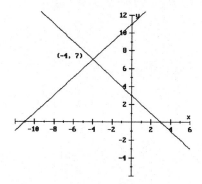

Since the slopes are different, this is an independent and consistent system. The graph shows the point of intersection.

21. l_1: $x = -y + 3$
$y = -x + 3$

l_2: $x = 3y - 1$
$x + 1 = 3y$
$\frac{1}{3}x + \frac{1}{3} = y$

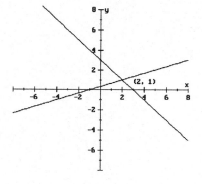

Since the slopes are different, this is an independent and consistent system. The graph shows the point of intersection.

25. l_1: $2x + 3y = -1$
$3y = -2x - 1$
$y = -\frac{2}{3}x - \frac{1}{3}$

l_2: $3x + 2y = 1$
$2y = -3x + 1$
$y = -\frac{3}{2}x + \frac{1}{2}$

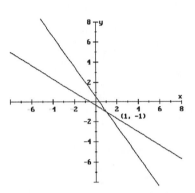

Since the slopes are different, this is an independent and consistent system. The graph shows the point of intersection.

29. l_1: $2x - y = 3$
$2x - 3 = y$

l_2: $y - 2x = 4$
$y = 2x + 4$

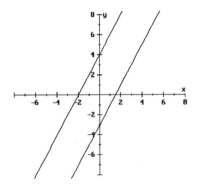

The slope of each line is 2. The y-intercept of l_1 is $(0, -3)$ and the y-intercept of l_2 is $(0, 4)$. Since the slopes are equal and the y-intercepts are different, the lines are parallel. The system has no solution.

33. l_1: $2x - 3y = 6$
$2x - 6 = 3y$
$\frac{2}{3}x - 2 = y$

l_2: $6y - 4x = 12$
$6y = 4x + 12$
$y = \frac{2}{3}x + 2$

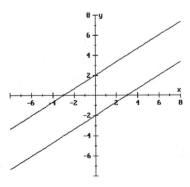

The slope of each line is $\frac{2}{3}$. The y-intercept of l_1 is $(0, -2)$ and the y-intercept of l_2 is $(0, 2)$. Since the slopes are equal and the y-intercepts are different, the lines are parallel. The system has no solution.

Exercises 7.1

37. l_1: $x - 2y = 5$
$x - 5 = 2y$
$\frac{1}{2}x - \frac{5}{2} = y$

l_2: $y = -3$

Since the slopes are different, this is an independent and consistent system. The graph shows the point of intersection.

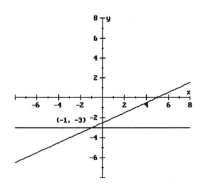

41. l_1: $x + y = 0$
$y = -x$

l_2: $2x + 4y = 1$
$4y = -2x + 1$
$y = -\frac{1}{2}x + \frac{1}{4}$

Since the slopes are different, this is an independent and consistent system. The graph shows the point of intersection.

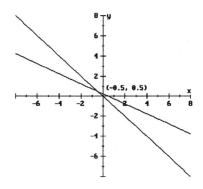

45. 1st number: x
2nd number: y

$x + y = 12$
$x - y = 2$

Sum is 12.
Difference is 2.

l_1: $x + y = 12$
$y = -x + 12$

l_2: $x - y = 2$
$x - 2 = y$

Since the slopes are different, this is an independent and consistent system. The graph shows the point of intersection.

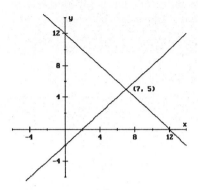

The numbers are 7 and 5.

49. One possible answer: The first is a consistent system. It has one solution. The second is an inconsistent system. It has no solution. The third is a dependent system and it has more than one solution.

53. l_1: $3x - 4y = 4$
$3x - 4 = 4y$
$\frac{3}{4}x - 1 = y$

l_2: $x + 2y = 8$
$2y = -x + 8$
$y = -\frac{1}{2}x + 4$

l_3: $y = \frac{5}{2}x - 8$

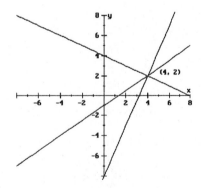

57. $16x^2 + 88x + 121$
$= (4x)^2 + 2(4x)(11) + (11)^2$
$= (4x + 11)^2$

Check to see that twice $4x$ and 11 is $88x$.

Exercises 7.2

61. $\dfrac{3x^2 - 7x + 2}{5x^2 - 3x - 14} = \dfrac{(3x - 1)(x - 2)}{(5x + 7)(x - 2)}$ Factor.

$\phantom{\dfrac{3x^2 - 7x + 2}{5x^2 - 3x - 14}} = \dfrac{3x - 1}{5x + 7}$ Reduce.

EXERCISES 7.2

1. $\begin{cases} y = x - 5 & (1) \\ x + y = 3 & (2) \end{cases}$

     $x + (x - 5) = 3$ Substitute $(x - 5)$ for y in Equation (2).
     $2x - 5 = 3$ Solve for x.
     $2x = 8$
     $x = 4$

     $y = x - 5$ Equation (1).
     $y = 4 - 5$ Substitute 4 for x.
     $y = -1$ Solve for y.

     $\{(4, -1)\}$

5. $\begin{cases} x = 11 - y & (1) \\ 2x - 3y = 12 & (2) \end{cases}$

     $2(11 - y) - 3y = 12$ Substitute $(11 - y)$ for x in Equation (2).
     $22 - 2y - 3y = 12$
     $22 - 5y = 12$ Solve for y.
     $-5y = -10$
     $y = 2$

     $x = 11 - y$ Equation (1).
     $x = 11 - 2$ Substitute 2 for y.
     $x = 9$ Solve for x.

     $\{(9, 2)\}$

9. $\begin{cases} x + y = -8 & (1) \\ -3x + 2y = 9 & (2) \end{cases}$

     $x + y = -8$
     $y = -x - 8$ Solve Equation (1) for y.
     $-3x + 2(-x - 8) = 9$ Substitute $(-x - 8)$ for y in Equation (2).
     $-3x - 2x - 16 = 9$ Solve for x.
     $-5x - 16 = 9$
     $-5x = 25$
     $x = -5$

     $x + y = -8$ Equation (1).
     $-5 + y = -8$ Substitute -5 for x.
     $y = -3$ Solve for y.

     $\{(-5, -3)\}$

13. $\begin{cases} x - y = 12 & (1) \\ 5y = 4x - 12 & (2) \end{cases}$

$$\begin{aligned} x - y &= 12 \\ x &= y + 12 \\ 5y &= 4(y + 12) - 12 \\ 5y &= 4y + 48 - 12 \\ 5y &= 4y + 36 \\ y &= 36 \end{aligned}$$

Solve Equation (1) for x.
Substitute $(y + 12)$ for x in Equation (2).

Solve for y.

$$\begin{aligned} x - y &= 12 \\ x - 36 &= 12 \\ x &= 48 \end{aligned}$$

Equation (1).
Substitute 36 for y.
Solve for x.

$\{(48, 36)\}$

17. $\begin{cases} x + 3y = 5 & (1) \\ 4x + 5y = 13 & (2) \end{cases}$

$$\begin{aligned} x + 3y &= 5 \\ x &= -3y + 5 \\ 4(-3y + 5) + 5y &= 13 \\ -12y + 20 + 5y &= 13 \\ -7y + 20 &= 13 \\ -7y &= -7 \\ y &= 1 \end{aligned}$$

Solve Equation (1) for x.
Substitute $(-3y + 5)$ for x in Equation (2).

Solve for y.

$$\begin{aligned} x + 3y &= 5 \\ x + 3(1) &= 5 \\ x + 3 &= 5 \\ x &= 2 \end{aligned}$$

Equation (1).
Substitute 1 for y.
Solve for x.

$\{(2, 1)\}$

21. $\begin{cases} 2a + 3b = 1 & (1) \\ a - 4b = 6 & (2) \end{cases}$

$$\begin{aligned} a - 4b &= 6 \\ a &= 4b + 6 \\ 2(4b + 6) + 3b &= 1 \\ 8b + 12 + 3b &= 1 \\ 11b + 12 &= 1 \\ 11b &= -11 \\ b &= -1 \end{aligned}$$

Solve Equation (2) for a.
Substitute $(4b + 6)$ for a in Equation (1).

Solve for b.

$$\begin{aligned} a - 4b &= 6 \\ a - 4(-1) &= 6 \\ a + 4 &= 6 \\ a &= 2 \end{aligned}$$

Equation (2).
Substitute -1 for b.
Solve for a.

$\{(2, -1)\}$

Exercises 7.2

25. $\begin{cases} 4x - y = 11 & (1) \\ -2x + 3y = -5 & (2) \end{cases}$

$$4x - y = 11$$
$$4x - 11 = y$$
$$-2x + 3(4x - 11) = -5$$
$$-2x + 12x - 33 = -5$$
$$10x - 33 = -5$$
$$10x = 28$$
$$x = \frac{14}{5}$$

Solve Equation (1) for y.
Substitute $(4x - 11)$ for y in Equation (2).

Solve for x.

$$4x - y = 11$$
$$4\left(\frac{14}{5}\right) - y = 11$$
$$\frac{56}{5} - y = 11$$
$$-y = -\frac{1}{5}$$
$$y = \frac{1}{5}$$

Equation (1).

Substitute $\frac{14}{5}$ for x.

Solve for y.

$\left\{\left(\frac{14}{5}, \frac{1}{5}\right)\right\}$

29. $\begin{cases} 2x + 3y = 23 & (1) \\ 2y - x = -1 & (2) \end{cases}$

$$2y - x = -1$$
$$2y + 1 = x$$
$$2(2y + 1) + 3y = 23$$
$$4y + 2 + 3y = 23$$
$$7y + 2 = 23$$
$$7y = 21$$
$$y = 3$$

Solve Equation (2) for x.
Substitute $(2y + 1)$ for x in Equation (1).

Solve for y.

$$2y - x = -1$$
$$2(3) - x = -1$$
$$6 - x = -1$$
$$-x = -7$$
$$x = 7$$

Equation (2).
Substitute 3 for y.
Solve for x.

$\{(7, 3)\}$

33. $\begin{cases} y = \frac{2}{3}x - 4 & (1) \\ 5y - x = 1 & (2) \end{cases}$

$$5\left(\frac{2}{3}x - 4\right) - x = 1$$

Substitute $\left(\frac{2}{3}x - 4\right)$ for y in Equation (2).

$$\frac{10}{3}x - 20 - x = 1$$

$$3\left(\frac{10}{3}x - 20 - x\right) = 3(1) \qquad \text{Solve for } x.$$

$$10x - 60 - 3x = 3$$
$$7x - 60 = 3$$
$$7x = 63$$
$$x = 9$$

$$y = \frac{2}{3}x - 4 \qquad \text{Equation (1).}$$

$$y = \frac{2}{3}(9) - 4 \qquad \text{Substitute 9 for } x.$$

$$y = 6 - 4 \qquad \text{Solve for } y.$$
$$y = 2$$

$\{(9, 2)\}$

37. $\begin{cases} x + 2y = \frac{3}{4} & (1) \\ 2x + y = -\frac{3}{8} & (2) \end{cases}$

$$x + 2y = \frac{3}{4}$$

$$x = -2y + \frac{3}{4} \qquad \text{Solve Equation (1) for } x.$$

$$2\left(-2y + \frac{3}{4}\right) + y = -\frac{3}{8} \qquad \text{Substitute } \left(-2y + \frac{3}{4}\right) \text{ for } x \text{ in Equation (2).}$$

$$-4y + \frac{3}{2} + y = -\frac{3}{8}$$

$$8\left(-4y + \frac{3}{2} + y\right) = 8\left(-\frac{3}{8}\right) \qquad \text{Multiply by 8 to clear the fractions.}$$

$$-32y + 12 + 8y = -3$$
$$-24y + 12 = -3 \qquad \text{Solve for } y.$$
$$-24y = -15$$

$$y = \frac{5}{8}$$

$$2x + y = -\frac{3}{8} \qquad \text{Equation (2).}$$

$$2x + \frac{5}{8} = -\frac{3}{8} \qquad \text{Substitute } \frac{5}{8} \text{ for } y.$$

$$2x = -1 \qquad \text{Solve for } x.$$

$$x = -\frac{1}{2}$$

$\left\{\left(-\frac{1}{2}, \frac{5}{8}\right)\right\}$

Exercises 7.2

41. $\begin{cases} x - 2y = 3 & (1) \\ 3x - 6y = 9 & (2) \end{cases}$

$$\begin{aligned} x - 2y &= 3 \\ x &= 2y + 3 \\ 3(2y + 3) - 6y &= 9 \\ 6y + 9 - 6y &= 9 \\ 9 &= 9 \end{aligned}$$

$\{(x, y) \mid x - 2y = 3\}$

Solve Equation (1) for x.
Substitute $(2y + 3)$ for x in Equation (2).

Identity.

45. $\begin{cases} \dfrac{x}{4} + \dfrac{y}{2} = \dfrac{7}{8} & (1) \\ 2x = 10 - 6y & (2) \end{cases}$

$$8\left(\dfrac{x}{4} + \dfrac{y}{2}\right) = 8\left(\dfrac{7}{8}\right)$$
$$2x + 4y = 7$$

Multiply Equation (1) by 8 to clear the fractions.
Call this Equation (1A).

$\begin{cases} 2x + 4y = 7 & (1A) \\ 2x = 10 - 6y & (2) \end{cases}$

$$\begin{aligned} 2x &= 10 - 6y \\ x &= 5 - 3y \\ 2(5 - 3y) + 4y &= 7 \\ 10 - 6y + 4y &= 7 \\ 10 - 2y &= 7 \\ -2y &= -3 \\ y &= \dfrac{3}{2} \end{aligned}$$

Solve Equation (2) for x.
Substitute $(5 - 3y)$ for x in Equation (1A).

Solve for y.

$$\begin{aligned} 2x + 4y &= 7 \\ 2x + 4\left(\dfrac{3}{2}\right) &= 7 \\ 2x + 6 &= 7 \\ 2x &= 1 \\ x &= \dfrac{1}{2} \end{aligned}$$

Equation (2).

Substitute $\dfrac{3}{2}$ for y.

Solve for x.

$\left\{\left(\dfrac{1}{2}, \dfrac{3}{2}\right)\right\}$

49. $\begin{cases} \dfrac{3}{4}x + 2y = 4 & (1) \\ 2x + 7y = 9 & (2) \end{cases}$

$$\dfrac{3}{4}x + 2y = 4$$

$$2y = -\dfrac{3}{4}x + 4 \qquad \text{Solve Equation (1) for } y.$$

$$y = -\dfrac{3}{8}x + 2$$

$$2x + 7\left(-\dfrac{3}{8}x + 2\right) = 9 \qquad \text{Substitute } \left(-\dfrac{3}{8}x + 2\right) \text{ for } y \text{ in Equation (2).}$$

$$2x - \dfrac{21}{8}x + 14 = 9$$

$$8\left(2x - \dfrac{21}{8}x + 14\right) = 8(9) \qquad \text{Solve for } x.$$

$$16x - 21x + 112 = 72$$
$$-5x + 112 = 72$$
$$-5x = -40$$
$$x = 8$$

$$\begin{aligned} 2x + 7y &= 9 \\ 2(8) + 7y &= 9 \\ 16 + 7y &= 9 \\ 7y &= -7 \\ y &= -1 \end{aligned} \qquad \begin{array}{l} \text{Equation (2).} \\ \text{Substitute 8 for } x. \\ \text{Solve for } y. \end{array}$$

$\{(8, -1)\}$

53. $\begin{cases} 0.3x - 0.1y = 0.9 & (1) \\ 0.5x - 2y = 7 & (2) \end{cases}$

$$\begin{aligned} 10(0.3x - 0.1y) &= 10(0.9) \\ 3x - y &= 9 \end{aligned} \qquad \begin{array}{l} \text{Multiply Equation (1) by 10 to clear the decimals.} \\ \text{Call this Equation (1A).} \end{array}$$

$$\begin{aligned} 10(0.5x - 2y) &= 10(7) \\ 5x - 20y &= 70 \end{aligned} \qquad \begin{array}{l} \text{Multiply Equation (2) by 10 to clear the decimals.} \\ \text{Call this Equation (2A).} \end{array}$$

$\begin{cases} 3x - y = 9 & (1A) \\ 5x - 20y = 70 & (2A) \end{cases}$

$$\begin{aligned} 3x - y &= 9 \\ 3x - 9 &= y \\ 5x - 20(3x - 9) &= 70 \\ 5x - 60x + 180 &= 70 \\ -55x + 180 &= 70 \\ -55x &= -110 \\ x &= 2 \end{aligned} \qquad \begin{array}{l} \text{Solve Equation (1A) for } y. \\ \text{Substitute } (3x - 9) \text{ for } y \text{ in Equation (2A.)} \\ \text{Solve for } x. \end{array}$$

Exercises 7.2

$$0.5x - 2y = 7$$
$$0.5(2) - 2y = 7$$
$$1 - 2y = 7$$
$$-2y = 6$$
$$y = -3$$

Equation (2).
Substitute 2 for x.
Solve for y.

$\{(2, -3)\}$

57.

	First Mix	Second Mix	Desired Mix
% of Bluegrass	0.40	0.70	0.55
Number of Pounds	x	y	600
Number of Pounds of Bluegrass Seed	$0.40x$	$0.70y$	$0.55(600) = 330$

$$\begin{cases} x + y = 600 & (1) \\ 0.40x + 0.70y = 330 & (2) \end{cases}$$

600 total pounds.
Information from chart.

$$10(0.40x + 0.70y) = 10(330)$$
$$4x + 7y = 3300$$

Multiply Equation (2) by 10 to clear the decimals. Call this Equation (2A).

$$\begin{cases} x + y = 600 & (1) \\ 4x + 7y = 3300 & (2A) \end{cases}$$

$$x + y = 600$$
$$x = -y + 600$$
$$4(-y + 600) + 7y = 3300$$
$$-4y + 2400 + 7y = 3300$$
$$2400 + 3y = 3300$$
$$3y = 900$$
$$y = 300$$

Solve Equation (1) for x.
Substitute $(-y + 600)$ for x in Equation (2A).

Solve for y.

$$x + y = 600$$
$$x + 300 = 600$$
$$x = 300$$

Equation (1).
Substitute 300 for y.
Solve for x.

300 lb of 40%; 300 lb of 70%

61.

	Rate	$\cdot$ Amount	= Income
1st Investment	0.12	x	$0.12x$
2nd Investment	0.15	y	$0.15y$

$$\begin{cases} x + y = 18000 & (1) \\ 0.12x = 0.15y & (2) \end{cases}$$

Total invested is 18000.
Incomes are equal.

$$100(0.12x) = 100(0.15y)$$
$$12x = 15y$$

Multiply Equation (2) by 100 to clear the decimals. Call this Equation (2A).

$$\begin{cases} x + y = 18000 & (1) \\ 12x = 15y & (2A) \end{cases}$$

$$x + y = 18000$$
$$x = -y + 18000$$
$$12(-y + 18000) = 15y$$
$$-12y + 216000 = 15y$$
$$216000 = 27y$$
$$8000 = y$$

Solve Equation (1) for x.
Substitute $(-y + 18000)$ for x in Equation (2A).

Solve for y.

$$x + y = 18000$$
$$x + 8000 = 18000$$
$$x = 10000$$

Equation (1).
Substitute 8000 for y.
Solve for x.

$10,000 at 12%; $8,000 at 15%

65.

	Rate	·	Amount	=	Interest
1st part	0.18		x		$0.18x$
2nd part	0.20		y		$0.20y$

$$\begin{cases} x + y = 2100 & (1) \\ 0.18x + 0.20y = 396 & (2) \end{cases}$$

Total borrowed is 2100.
Sum of the interests is 396.

$$100(0.18x + 0.20y) = 100(396)$$
$$18x + 20y = 39600$$

Multiply Equation (2) by 100 to clear the decimals.
Call this Equation (2A).

$$\begin{cases} x + y = 2100 & (1) \\ 18x + 20y = 39600 & (2A) \end{cases}$$

$$x + y = 2100$$
$$x = -y + 2100$$
$$18(-y + 2100) + 20y = 39600$$
$$-18y + 37800 + 20y = 39600$$
$$2y + 37800 = 39600$$
$$2y = 1800$$
$$y = 900$$

Solve Equation (1) for x.
Substitute $(-y + 2100)$ for x in Equation (2A).

Solve for y.

$$x + y = 2100$$
$$x + 900 = 2100$$
$$x = 1200$$

Equation (1).
Substitute 900 for y.
Solve for x.

$1200 at 18%; $900 at 20%

69. $\begin{cases} 3x + 5y = 7 & (1) \\ 4x - 2y = 12 & (2) \end{cases}$

$$4x - 2y = 12$$
$$4x - 12 = 2y$$
$$2x - 6 = y$$
$$3x + 5(2x - 6) = 7$$
$$3x + 10x - 30 = 7$$
$$13x - 30 = 7$$
$$13x = 37$$

$$x = \frac{37}{13}$$

Solve Equation (2) for y.

Substitute $(2x - 6)$ for y in Equation (1).

Solve for x.

Exercises 7.2

$$4x - 2y = 12 \qquad \text{Equation (2).}$$

$$4\left(\frac{37}{13}\right) - 2y = 12 \qquad \text{Substitute } \frac{37}{13} \text{ for } x.$$

$$\frac{148}{13} - 2y = 12$$

$$13\left(\frac{148}{13} - 2y\right) = 13(12) \qquad \text{Multiply both sides by 13 to clear the fractions.}$$

$$148 - 26y = 156 \qquad \text{Solve for } y.$$
$$-26y = 8$$

$$y = -\frac{4}{13}$$

$$\left\{\left(\frac{37}{13}, -\frac{4}{13}\right)\right\}$$

73. $9x - 4y = 36$

y-intercept:
$$9(0) - 4y = 36 \qquad \text{Replace } x \text{ with 0.}$$
$$-4y = 36$$
$$y = -9 \qquad (0, -9).$$

x-intercept:
$$9x - 4(0) = 36 \qquad \text{Replace } y \text{ with 0.}$$
$$9x = 36$$
$$x = 4 \qquad (4, 0).$$

77. $m = \dfrac{y_2 - y_1}{x_2 - x_1}$ \qquad Slope formula.

$m = \dfrac{-1 - (-2)}{-2 - (-5)}$ \qquad Substitute: (x_1, y_1) is $(-5, -2)$ and (x_2, y_2) is $(-2, -1)$.

$m = \dfrac{1}{3}$ \qquad Simplify.

81. $x = -\dfrac{4}{3}y + 3$

$3x = -4y + 9$ \qquad Multiply both sides by 3.
$4y = -3x + 9$ \qquad Solve for y.

$y = -\dfrac{3}{4}x + \dfrac{9}{4}$ \qquad $y = mx + b$.

$m = -\dfrac{3}{4}$ \qquad Slope.

EXERCISES 7.3

1. $\begin{cases} x + y = 4 \\ x - y = 2 \end{cases}$

   $2x = 6$ Add the two equations.
   $x = 3$ Solve for x.

   $3 + y = 4$ Substitute 3 for x in Equation (1).
   $y = 1$ Solve for y.

   $\{(3, 1)\}$

5. $\begin{cases} 2x - y = 3 \\ -2x + y = 2 \end{cases}$

   $0 = 5$ This is a false statement so there is no solution.

   $\emptyset$

9. $\begin{cases} x + 3y = 10 \\ x - y = -2 \end{cases}$

   $\begin{cases} x + 3y = 10 \\ -x + y = 2 \end{cases}$ Multiply Equation (2) by -1.

   $4y = 12$ Add the two equations.
   $y = 3$ Solve for y.

   $x - 3 = -2$ Substitute 3 for y in Equation (2).
   $x = 1$ Solve for x.

   $\{(1, 3)\}$

13. $\begin{cases} 2x + y = 5 \\ 4x + y = 6 \end{cases}$

    $\begin{cases} -2x - y = -5 \\ 4x + y = 6 \end{cases}$ Multiply Equation (1) by -1.

    $2x = 1$ Add the two equations.

    $x = \dfrac{1}{2}$ Solve for x.

    $2\left(\dfrac{1}{2}\right) + y = 5$ Substitute $\dfrac{1}{2}$ for x in Equation (1).
    $1 + y = 5$
    $y = 4$ Solve for y.

    $\left\{\left(\dfrac{1}{2}, 4\right)\right\}$

Exercises 7.3

17. $\begin{cases} 3x - 2y = -11 \\ x + 2y = -1 \end{cases}$

$\quad\quad\quad 4x = -12$ Add the two equations.
$\quad\quad\quad\ \ x = -3$ Solve for x.

$\quad -3 + 2y = -1$ Substitute -3 for x in Equation (2).
$\quad\quad\quad\ 2y = 2$
$\quad\quad\quad\ \ y = 1$ Solve for y.

$\{(-3, 1)\}$

21. $\begin{cases} x + 5y + 1 = 0 \\ 2x + 7y - 1 = 0 \end{cases}$

$\begin{cases} x + 5y = -1 \\ 2x + 7y = 1 \end{cases}$ Get each equation in the form $Ax + By = C$.

$\begin{cases} -2x - 10y = 2 \\ \ \ 2x + 7y = 1 \end{cases}$ Multiply Equation (1A) by -2.

$\quad\quad\quad -3y = 3$ Add the two equations.
$\quad\quad\quad\ \ \ y = -1$ Solve for y.

$x + 5(-1) + 1 = 0$ Substitute -1 for y in Equation (1).
$\quad\ x - 5 + 1 = 0$
$\quad\quad\ \ x - 4 = 0$ Solve for x.
$\quad\quad\quad\ \ \ x = 4$

$\{(4, -1)\}$

25. $\begin{cases} 2x + 4y = 5 \\ 3x + 6y = 6 \end{cases}$

$\begin{cases} -6x - 12 = -15 \\ \ \ 6x + 12y = 12 \end{cases}$ Multiply Equation (1) by -3.
  Multiply Equation (2) by 2.

$\quad\quad\quad\ \ 0 = -3$ Add the two equations.
  This is a false statement so there is no solution.

$\emptyset$

29. $\begin{cases} 2x - 2y = 3 \\ 15x - 4y = 0 \end{cases}$

$\begin{cases} -4x + 4y = -6 \\ \ 15x - 4y = 0 \end{cases}$ Multiply Equation (1) by -2.

$\quad\quad\quad 11x = -6$ Add the two equations.

$\quad\quad\quad\ \ x = -\dfrac{6}{11}$ Solve for x.

$2\left(-\dfrac{6}{11}\right) - 2y = 3$ Substitute $-\dfrac{6}{11}$ for x in Equation (1).

$\quad -\dfrac{12}{11} - 2y = 3$

$$-12 - 22y = 33$$
$$-22y = 45$$

Multiply both sides by 11.
Solve for y.

$$y = -\frac{45}{22}$$

$$\left\{\left(-\frac{6}{11}, -\frac{45}{22}\right)\right\}$$

33. $\begin{cases} 4x + 9y = 57 \\ 9x - 4y = -90 \end{cases}$

$\begin{cases} 16x + 36y = 228 \\ 81x - 36y = -810 \end{cases}$

Multiply Equation (1) by 4.
Multiply Equation (2) by 9.

$$97x = -582$$
$$x = -6$$

Add the two equations.
Solve for x.

$$4(-6) + 9y = 57$$

Substitute -6 for x in Equation (1).

$$-24 + 9y = 57$$
$$9y = 81$$
$$y = 9$$

Solve for y.

$\{(-6, 9)\}$

37. $\begin{cases} 3x = 5y + 10 \\ 7x - 3y - 14 = 0 \end{cases}$

$\begin{cases} 3x - 5y = 10 \\ 7x - 3y = 14 \end{cases}$

Write each equation in the form $Ax + By = C$.

$\begin{cases} 9x - 15y = 30 \\ -35x + 15y = -70 \end{cases}$

Multiply Equation (1A) by 3.
Multiply Equation (2A) by -5.

$$-26x = -40$$

Add the two equations.

$$x = \frac{20}{13}$$

Solve for x.

$$3\left(\frac{20}{13}\right) = 5y + 10$$

Substitute $\frac{20}{13}$ for x in Equation (1).

$$\frac{60}{13} = 5y + 10$$

$$60 = 65y + 130$$
$$-70 = 65y$$

Multiply both sides by 13.
Solve for y.

$$-\frac{14}{13} = y$$

$\left\{\left(\frac{20}{13}, -\frac{14}{13}\right)\right\}$

Exercises 7.3

41. $\begin{cases} 2x + y + 2 = 0 \\ 5x = y + 23 \end{cases}$

$\begin{cases} 2x + y = -2 \\ 5x - y = 23 \end{cases}$ Write each equation in the form $Ax + By = C$.

$7x = 21$ Add the two equations.
$x = 3$ Solve for x.

$5(3) = y + 23$ Substitute 3 for x in Equation (2).
$15 = y + 23$
$-8 = y$ Solve for y.

$\{(3, -8)\}$

45. $\begin{cases} \dfrac{1}{2}x + \dfrac{2}{3}y = \dfrac{7}{3} \\ 3x - 2y = -16 \end{cases}$

$6\left(\dfrac{1}{2}x + \dfrac{2}{3}y\right) = 6\left(\dfrac{7}{3}\right)$ Multiply Equation (1) by 6 to clear the fractions.

$3x + 4y = 14$

$\begin{cases} 3x + 4y = 14 \\ 3x - 2y = -16 \end{cases}$

$\begin{cases} 3x + 4y = 14 \\ -3x + 2y = 16 \end{cases}$ Multiply Equation (2) by -1.

$6y = 30$ Add the two equations.
$y = 5$ Solve for y.

$3x - 2(5) = -16$ Substitute 5 for y in Equation (2).
$3x - 10 = -16$
$3x = -6$ Solve for x.
$x = -2$

$\{(-2, 5)\}$

49. $\begin{cases} 9x + 16y = -65 \\ 12x + 11y = -4 \end{cases}$

$\begin{cases} -36x - 64y = 260 \\ -36x + 33y = -12 \end{cases}$ Multiply Equation (1) by -4.
     Multiply Equation (2) by 3.

$31y = 248$ Add the two equations.
$y = -8$ Solve for y.

$12x + 11(-8) = -4$ Substitute -8 for y in Equation (2).
$12x - 88 = -4$
$12x = 84$ Solve for x.
$x = 7$

$\{(7, -8)\}$

53. $\begin{cases} 0.5x + 0.7y = 0.1 \\ 7x + 0.5y = -0.1 \end{cases}$

$\begin{cases} 5x + 7y = 1 \\ 7x + 5y = -1 \end{cases}$ Multiply both equations by 10 to clear the decimals.

$\begin{cases} -25x - 35y = -5 \\ 49x + 35y = -7 \end{cases}$ Multiply Equation (1A) by -5.
Multiply Equation (2A) by 7.

$\qquad 24x = -12$
$\qquad\ \ x = -0.5$ Add the two equations. Solve for x.

$0.5(-0.5) + 0.7y = 0.1$
$\quad -0.25 + 0.7y = 0.1$ Substitute -0.5 for x in Equation (1).
$\qquad\quad 0.7y = 0.35$
$\qquad\qquad y = 0.5$ Solve for y.

$\{(-0.5, 0.5)\}$

57.

	% pure gold	· oz	=	Number of oz of pure gold
1st gold	0.85	x		$0.85x$
2nd gold	0.70	y		$0.70y$
Mixture	0.75	15		$0.75(15) = 11.25$

$\begin{cases} x + y = 15 \\ 0.85x + 0.70y = 11.25 \end{cases}$ Total ounces is 15.
1st gold + 2nd gold = desired gold mixture.

$\begin{cases} x + y = 15 \\ 85x + 70y = 1125 \end{cases}$ Multiply Equation (2) by 100 to clear the decimals.

$\begin{cases} -70x - 70y = -1050 \\ 85x + 70y = 1125 \end{cases}$ Multiply Equation (1) by -70.

$\qquad 15x = 75$
$\qquad\ \ x = 5$ Add the two equations. Solve for x.

$5 + y = 15$
$\qquad y = 10$ Substitute 5 for x in Equation (1). Solve for y.

5 oz of 85%; 10 oz of 70%

61.

	price	· number	=	total sales
1 year	27	x		$27x$
2 year	48	y		$48y$

Exercises 7.3

$$\begin{cases} x + y = 210 \\ 27x + 48y = 6930 \end{cases}$$ Number of subscriptions equals 210.
Total sales is 6930.

$$\begin{cases} -27x - 27y = -5670 \\ 27x + 48y = 6930 \end{cases}$$ Multiply Equation (1) by -27.

$$\begin{aligned} 21y &= 1260 \\ y &= 60 \end{aligned}$$ Add the two equations.
Solve for y.

$$\begin{aligned} x + 60 &= 210 \\ x &= 150 \end{aligned}$$ Substitute 60 for y in Equation (1).

150 @ \$27; 60 @ \$48

65. One possible answer: First make sure each equation is in the standard form $Ax + By = C$. Next multiply one or both equations by the appropriate factors so that either the x-coefficients or the y-coefficients are opposites. Then add the two equations and one of the variables will be eliminated. Solve for the remaining variable. Substitute the value found for the remaining variable into one of the original equations and solve for the other variable. Write the solution set.

69. $ax + by = 1$

$$\begin{aligned} a(2) + b(3) &= 1 \\ a(-3) + b(-5) &= 1 \end{aligned}$$ Substitute 2 for x and 3 for y.
Substitute -3 for x and -5 for y.
The coordinates of the solutions must satisfy the equation.

$$\begin{cases} 2a + 3b = 1 \\ -3a - 5b = 1 \end{cases}$$

$$\begin{cases} 6a + 9b = 3 \\ -6a - 10b = 2 \end{cases}$$ Multiply Equation (1) by 3.
Multiply Equation (2) by 2.

$$\begin{aligned} -b &= 5 \\ b &= -5 \end{aligned}$$ Add the two equations.
Solve for b.

$$\begin{aligned} 2a + 3(-5) &= 1 \\ 2a - 15 &= 1 \\ 2a &= 16 \\ a &= 8 \end{aligned}$$ Substitute -5 for b in Equation (1).
Solve for a.

$8x - 5y = 1$ Substitute 8 for a and -5 for b.

73. $4x^2 - 1 = 0$
     $(2x + 1)(2x - 1) = 0$ Factor the left side.
     $2x + 1 = 0$ or $2x - 1 = 0$ Zero-product property.
         $2x = -1$ or $2x = 1$
         $x = -\dfrac{1}{2}$ or $x = \dfrac{1}{2}$

$\left\{ -\dfrac{1}{2}, \dfrac{1}{2} \right\}$

77. $\dfrac{x^2 - 1}{16a^3} \cdot \dfrac{12a}{x + 1}$

$= \dfrac{(x + 1)(x - 1)}{16a^3} \cdot \dfrac{12a}{x + 1}$ Factor.

$= \dfrac{x - 1}{4a^2} \cdot \dfrac{3}{1}$ Reduce.

$= \dfrac{3x - 3}{4a^2}$ Multiply.

EXERCISES 7.4

1. $\begin{cases} D = 30000 + 12x \\ D = 16x \end{cases}$

   $16x = 30000 + 12x$ Substitute $16x$ for D in Equation (1).
   $4x = 30000$
   $x = 7500$ Solve for x.

   7500 changes

5. sales: x
   income: y

   $\begin{cases} y = 0.05x \\ y = 180 + 0.02x \end{cases}$ Income at 1st store.
                                               Income at 2nd store.

   $0.05x = 180 + 0.02x$ Substitute $0.05x$ for y in Equation (2).
   $0.03x = 180$
   $x = 6000$ Solve for x.

   $y = 0.05(6000)$ Substitute 6000 for x in Equation (1).
   $y = 300$ Solve for y.

   Sales: $6000; Income: $300

9.
	price ·	number =	sales
1st ticket	25	x	$25x$
2nd ticket	15	y	$15y$

   $\begin{cases} x + y = 2600 \\ 25x + 15y = 46000 \end{cases}$ 2600 tickets are sold.
                                Total sales of 46000.

   $\begin{cases} -15x - 15y = -39000 \\ 25x + 15y = 46000 \end{cases}$ Multiply Equation (1) by -15.

   $10x = 7000$ Add the two equations.
   $x = 700$ Solve for x.

   $700 + y = 2600$ Substitute 700 for x in Equation (1).
   $y = 1900$ Solve for y.

   700 @ $25; $1900 @ $15

Exercises 7.4

13. One possible answer: The graphing method is the least desirable since it is hard to recognize the exact solution if it contains fractions. It also takes longer than the other two methods. If one of the equations is written as one of the variables in terms of the other then the substitution method is the easiest. If the equations are in standard form then the elimination method is easiest. In general, the method of elimination is the easiest to use with 3 variables.

Example:

$$\begin{cases} x + 2y - z = 2 \\ 2x + y + z = 7 \\ -3x - y + z = -2 \end{cases}$$

$\begin{array}{l} x + 2y - z = 2 \\ \underline{2x + y + z = 7} \\ 3x + 3y = 9 \end{array}$ Add Equations (1) and (2).

Call this Equation (4).

$\begin{array}{l} x + 2y - z = 2 \\ \underline{-x - y + z = -2} \\ -2x + y = 0 \end{array}$ Add Equations (1) and (3). Eliminate the same variable that was eliminated in the first addition. Call this Equation (5).

$\begin{array}{l} 3x + 3y = 0 \\ \underline{6x - 3y = 0} \\ 9x = 9 \\ x = 1 \end{array}$ Equation (4). Multiply Equation (5) by -3.

Add the equations and solve for x.

$\begin{array}{r} -2(1) + y = 0 \\ -2 + y = 0 \\ y = 2 \end{array}$ Substitute 1 for x in Equation (5).

Solve for y.

$\begin{array}{r} 2(1) + 2 + z = 7 \\ 2 + 2 + z = 7 \\ 4 + z = 7 \\ z = 3 \end{array}$ Substitute 1 for x and 2 for y in Equation (2).

Solve for z.

$\{(1, 2, 3)\}$

17. $3x^2 - 34x - 24$
    $= 3x^2 - 36x + 2x - 24$ $mn = 3(-24) = -72;\ m + n = -34$
                                          $(-36)(2) = -72;\ -36 + 2 = -34$
    $= 3x(x - 12) + 2(x - 12)$ Rewrite the trinomial.
    $= (x - 12)(3x + 2)$ Factor by grouping.

21. $64x^2 - 100$
    $= 4(16x^2 - 25)$ The GCF is 4.
    $= 4[(4x)^2 - (5)^2]$ Write $16x$ as the square of $4x$ and $16x^2$ as the square of $4x$ and 25 as the square of 5.

    $= 4(4x + 5)(4x - 5)$ Factor the difference of two squares.

25. $\dfrac{2a^2 + 7ab + 3b^2}{2a^2 + 5ab - 3b^2}$

$= \dfrac{(2a + b)(a + 3b)}{(2a - b)(a + 3b)}$ Factor.

$= \dfrac{2a + b}{2a - b}$ Reduce.

CHAPTER 7 CONCEPT REVIEW

1. True

2. False;   It has exactly one solution.

3. True

4. True

5. False;   It may contain more than two variables.

6. True

7. True

8. False;   An identity indicates that there are an infinite number of solutions, but not all pairs are solutions.

9. True

10. False;   It can be a solution.

CHAPTER 7 TEST

1. $\begin{cases} x = 2y \\ x - y = 3 \end{cases}$

     $(2y) - y = 3$ Substitute $(2y)$ for x in Equation (2).
     $y = 3$

     $x = 2y$ Equation (1).
     $x = 2(3)$ Substitute 3 for y.
     $x = 6$

     $\{(6, 3)\}$

2. $\begin{cases} 2x + 3y = 5 \\ x - y = 7 \end{cases}$

     $\begin{cases} 2x + 3y = 5 \\ 3x - 3y = 21 \end{cases}$ Multiply Equation (2) by 3.

     $5x = 26$ Add the two equations.

     $x = \dfrac{26}{5}$ Solve for x.

Chapter 7 Test

$$\frac{26}{5} - y = 7$$

Substitute $\frac{26}{5}$ for x in Equation (2).

$$26 - 5y = 35$$
$$-5y = 9$$

$$y = -\frac{9}{5}$$

Solve for y.

$$\left\{\left(\frac{26}{5}, -\frac{9}{5}\right)\right\}$$

3. l_1: $y = -x - 10$

   l_2: $y = 2x + 2$

Since the slopes are different, this is an independent and consistent system. The graph shows the point of intersection.

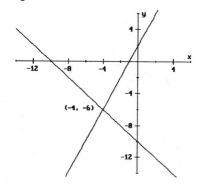

4. $\begin{cases} y = 2x + 3 \\ 3x - 2y = 7 \end{cases}$

$$3x - 2(2x + 3) = 7$$

Substitute $(2x + 3)$ for y in Equation (2).

$$3x - 4x - 6 = 7$$
$$-x - 6 = 7$$
$$-x = 13$$
$$x = -13$$

Solve for x.

$$y = 2(-13) + 3$$
$$y = -26 + 3$$
$$y = -23$$

Substitute -13 for x in Equation (1).

Solve for y.

$\{(-13, -23)\}$

5. $\begin{cases} x - 2y = -2 \\ 3x - y = 20 \end{cases}$

$$x - 2y = -2$$
$$x = 2y - 2$$
$$3(2y - 2) - y = 20$$

Solve Equation (1) for x.
Substitute $(2y - 2)$ for x in Equation (2).

$$6y - 6 - y = 20$$
$$5y - 6 = 20$$
$$5y = 26$$

Solve for y.

$$y = \frac{26}{5}$$

$$x - 2\left(\frac{26}{5}\right) = -2$$ Substitute $\frac{26}{5}$ for y in Equation (1).

$$x - \frac{52}{5} = -2$$ Solve for x.

$$5x - 52 = -10$$
$$5x = 42$$

$$x = \frac{42}{5}$$

$$\left\{\left(\frac{42}{5}, \frac{26}{5}\right)\right\}$$

6. $\begin{cases} 5x - 4y = 20 \\ 3x + 2y = -10 \end{cases}$

$\begin{cases} 5x - 4y = 20 \\ 6x + 4y = -20 \end{cases}$ Multiply Equation (2) by 2.

$$11x = 0$$ Add the two equations.
$$x = 0$$ Solve for x.

$$3(0) + 2y = -10$$ Substitute 0 for x in Equation (2).
$$2y = -10$$ Solve for y.
$$y = -5$$

$\{(0, -5)\}$

7. 1st number: x
2nd number: y

$\begin{cases} y + 3x = 7 \\ 3y - x = 1 \end{cases}$ $y + 3x$: Three times the first added to the second.
$3y - x$: First subtracted from three times the second.

$\begin{cases} y + 3x = 7 \\ 9y - 3x = 3 \end{cases}$ Multiply Equation (2) by 3.

$$10y = 10$$ Add the two equations.
$$y = 1$$ Solve for y.

$$1 + 3x = 7$$ Substitute 1 for y in Equation (1).
$$3x = 6$$
$$x = 2$$

The numbers are 2 and 1.

8.

	1st Mix	· 2nd Mix	= Desired Mix
% of Bluegrass	0.45	0.60	0.50
Number of Pounds	x	y	15
Number of Pounds of Bluegrass Seed	$0.45x$	$0.60y$	$0.50(15) = 7.5$

$$\begin{cases} x + y = 15 \\ 0.45x + 0.60y = 7.5 \end{cases}$$ 15 total pounds. Information from chart.

$$\begin{cases} x + y = 15 \\ 45x + 60y = 750 \end{cases}$$ Multiply both sides by 100.

$$\begin{cases} -45x - 45y = -675 \\ 45x + 60y = 750 \end{cases}$$ Multiply Equation (1) by -45.

$15y = 75$
$y = 5$

Add the two equations.
Solve for y.

$x + 5 = 15$
$x = 10$

Substitute 5 for y in Equation (1).

10 lb @ 45%; 5 lb @ 60%

CHAPTER 8

EXERCISES 8.1

1. $\sqrt{4} = 2$ The root is 2 because $2 \cdot 2 = 2^2 = 4$.

5. $-\sqrt{81} = -9$ $-\sqrt{81} = -(9) = -9$.

9. $\sqrt{\dfrac{9}{25}} = \dfrac{3}{5}$ The root is $\dfrac{3}{5}$ because $\left(\dfrac{3}{5}\right)\left(\dfrac{3}{5}\right) = \dfrac{9}{25}$.

13. $\sqrt{0.64} = \sqrt{\dfrac{64}{100}}$ $0.64 = \dfrac{64}{100}$.

    $= \dfrac{8}{10}$ The root is $\dfrac{8}{10}$ because $\left(\dfrac{8}{10}\right)\left(\dfrac{8}{10}\right) = \dfrac{64}{100}$.

    $= 0.8$

17. $\sqrt{676} = 26$ Calculator.

21. $-\sqrt{961} = -31$ Calculator.

25. $\sqrt{\dfrac{121}{225}} = \dfrac{11}{15}$ Calculator. $\sqrt{121} = 11$ and $\sqrt{225} = 15$.

29. $-\sqrt{6.67} = -2.6$ Calculator.

33. $\sqrt{14400} = 120$ Calculator.

37. $\sqrt{7} \approx 2.65$ Calculator.

41. $\sqrt{200} \approx 14.14$ Calculator.

45. $-\sqrt{892} \approx -29.87$ Calculator.

49. $\sqrt{14.4} \approx 3.79$ Calculator.

53. $\sqrt{109.2} \approx 10.45$ Calculator.

57. $v = \sqrt{2.5r}$ Formula.
    $v = \sqrt{2.5(250)}$ Replace r with 250.
    $v = \sqrt{625}$ Multiply.
    $v = 25$ $25 \cdot 25 = 625$.

    25 mph

Exercises 8.1

61.
$$l\sqrt{s} = k \quad \text{Formula.}$$

$$16\sqrt{\frac{1}{64}} = k \quad \text{Replace } l \text{ with 16 and } s \text{ with } \frac{1}{64}.$$

$$16\left(\frac{1}{8}\right) = k \quad \text{Solve for } k. \quad \frac{1}{8} \cdot \frac{1}{8} = \frac{1}{64}.$$

$$2 = k$$

$$l\sqrt{s} = 2 \quad \text{Formula with 2 substituted for } k.$$

$$l\sqrt{\frac{1}{16}} = 2 \quad \text{Replace } s \text{ with } \frac{1}{16}.$$

$$l\left(\frac{1}{4}\right) = 2 \quad \frac{1}{4} \cdot \frac{1}{4} = \frac{1}{16}.$$

$$l = 8 \quad \text{Solve for } l.$$

65. One possible answer: To check the answer, square 10.72 and see if the result is approximately 115.

$$(10.72)(10.72) = 114.9 \approx 115$$

69. $22^2 = 484$
    $23^2 = 529$

    $22 < \sqrt{519} < 23$

73. $B(2, 5)$ — Two units right and five units up.

77. $2y - 5x = 10$

x	y
0	5
-2	0
-4	-5

One point can be found by letting $x = 0$, another by letting $y = 0$ and a third by letting $x = -4$. Plot these three points and connect them with a straight line.

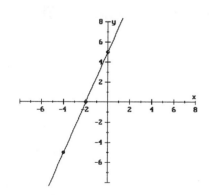

81. $6x - 12y = 7$

x	y
0	-0.6
1.2	0
3.2	1

One point can be found by letting $x = 0$, another by letting $y = 0$ and a third by letting $y = 1$. Plot these three points and connect them with a straight line.

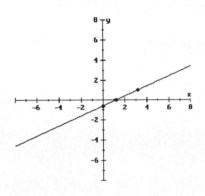

EXERCISES 8.2

1. $\sqrt{8} = \sqrt{4} \cdot \sqrt{2}$
   $= 2\sqrt{2}$

   The largest square factor of 8 is 4, so write $\sqrt{4} \cdot \sqrt{2}$.

5. $\sqrt{40} = \sqrt{4} \cdot \sqrt{10}$
   $= 2\sqrt{10}$

   The largest square factor of 40 is 4, so write $\sqrt{4} \cdot \sqrt{10}$.

9. $\sqrt{16c^2} = \sqrt{16} \cdot \sqrt{c^2}$
   $= 4c$

   $c \cdot c = c^2$.

13. $\sqrt{32x^2y^2} = \sqrt{16} \cdot \sqrt{x^2} \cdot \sqrt{y^2} \cdot \sqrt{2}$
    $= 4xy\sqrt{2}$

    $\sqrt{32} = \sqrt{16} \cdot \sqrt{2}$.

17. $\sqrt{80} = \sqrt{16} \cdot \sqrt{5}$
    $= 4\sqrt{5}$

    The largest square factor of 80 is 16, so write $\sqrt{16} \cdot \sqrt{5}$.

21. $\sqrt{44} = \sqrt{4} \cdot \sqrt{11}$
    $= 2\sqrt{11}$

    The largest square factor of 44 is 4, so write $\sqrt{4} \cdot \sqrt{11}$.

25. $\sqrt{72x^2} = \sqrt{36} \cdot \sqrt{x^2} \cdot \sqrt{2}$
    $= 6x\sqrt{2}$

    $\sqrt{72} = \sqrt{36} \cdot \sqrt{2}$.

29. $\sqrt{288x^2y^4} = \sqrt{144} \cdot \sqrt{x^2} \cdot \sqrt{y^4} \cdot \sqrt{2}$
    $= 12xy^2\sqrt{2}$

    $\sqrt{288} = \sqrt{144} \cdot \sqrt{2}$.
    $y^2 \cdot y^2 = y^4$.

33. $\sqrt{200x^4y^3} = \sqrt{100} \cdot \sqrt{x^4} \cdot \sqrt{y^2} \cdot \sqrt{2} \cdot \sqrt{y}$
    $= 10x^2y\sqrt{2y}$

    $\sqrt{200} = \sqrt{100} \cdot \sqrt{2}$.
    $\sqrt{y^3} = \sqrt{y^2} \cdot \sqrt{y}$.
    $x^2 \cdot x^2 = x^4$.

Exercises 8.2

37. $\sqrt{192} = \sqrt{64} \cdot \sqrt{3}$ The largest square factor of 192 is 64, so write $\sqrt{64} \cdot \sqrt{3}$.
    $= 8\sqrt{3}$

41. $\sqrt{198w^6} = \sqrt{9} \cdot \sqrt{w^6} \cdot \sqrt{22}$ $\sqrt{198} = \sqrt{9} \cdot \sqrt{22}$.
    $= 3w^3\sqrt{22}$ $w^3 \cdot w^3 = w^6$.

45. $\sqrt{156a^4} = \sqrt{4} \cdot \sqrt{a^4} \cdot \sqrt{39}$ $\sqrt{156} = \sqrt{4} \cdot \sqrt{39}$.
    $= 2a^2\sqrt{39}$ $a^2 \cdot a^2 = a^4$.

49. $\sqrt{32p^8} = \sqrt{16} \cdot \sqrt{p^8} \cdot \sqrt{2}$ $\sqrt{32} = \sqrt{16} \cdot \sqrt{2}$.
    $= 4p^4\sqrt{2}$ $p^4 \cdot p^4 = p^8$.

53. $\sqrt{1125r^9s^5} = \sqrt{225} \cdot \sqrt{r^8} \cdot \sqrt{s^4} \cdot \sqrt{5} \cdot \sqrt{r}\sqrt{s}$ $\sqrt{1125} = \sqrt{225} \cdot \sqrt{5}$.
                                                   $\sqrt{r^9} = \sqrt{r^8} \cdot \sqrt{r}$.
                                                   $\sqrt{s^5} = \sqrt{s^4} \cdot \sqrt{s}$.
                                                   $r^4 \cdot r^4 = r^8$.
                  $= 15r^4s^2\sqrt{5rs}$ $s^2 \cdot s^2 = s^4$.

57. side length: x

     Area = (side)^2 Formula.
     $4500 = x^2$ Area = 4500.
     $\sqrt{4500} = x$

     $4500 = 2^2 \cdot 3^2 \cdot 5^3$ The prime factors of 4500.

     $\sqrt{4500} = \sqrt{2^2} \cdot \sqrt{3^2} \cdot \sqrt{5^2} \cdot \sqrt{5}$
               $= 2 \cdot 3 \cdot 5\sqrt{5}$
               $= 30\sqrt{5}$

     $30\sqrt{5}$ ft ≈ 67.1 ft

61. $V = \sqrt{ER}$ Formula.
     $V = \sqrt{600 \cdot 18}$ Replace E with 600 and R with 18.
     $V = \sqrt{6 \cdot 100 \cdot 6 \cdot 3}$ $600 = 6 \cdot 100$ and $18 = 6 \cdot 3$.
     $V = \sqrt{6^2} \cdot \sqrt{100} \cdot \sqrt{3}$
     $V = 6 \cdot 10\sqrt{3} = 60\sqrt{3}$

     $60\sqrt{3}$ volts

64. One possible answer: In the Product Property of Radicals it is specified that the factors a and b are non-negative numbers. We cannot factor -18 with non-negative numbers only.

69. $\sqrt{x^2 + 4x + 4} = \sqrt{(x + 2)^2}$ Factor the radicand.
                     $= x + 2$ $(x + 2) \cdot (x + 2) = (x + 2)^2$.

73. $3[2 - 3(x + 5)] - 18 = 0$
    $3(2 - 3x - 15) - 18 = 0$ — Multiply inside the brackets.
    $3(-3x - 13) - 18 = 0$ — Combine like terms inside the parentheses.
    $-9x - 39 - 18 = 0$ — Multiply.
    $-9x - 57 = 0$ — Combine like terms.
    $-9x = 57$ — Solve for x.
    $x = -\dfrac{19}{3}$

    $\left\{-\dfrac{19}{3}\right\}$

77. $(x + 7)^2 - 4(x + 3)(2x - 7) = (x - 1)^2$
    $x^2 + 14x + 49 - 4(2x^2 - x - 21) = x^2 - 2x + 1$ — Multiply using FOIL.
    $x^2 + 14x + 49 - 8x^2 + 4x + 84 = x^2 - 2x + 1$ — Multiply.
    $-7x^2 + 18x + 133 = x^2 - 2x + 1$ — Combine like terms.
    $0 = 8x^2 - 20x - 132$
    $0 = 2x^2 - 5x - 33$ — Multiply both sides by $\dfrac{1}{4}$.
    $0 = (2x - 11)(x + 3)$ — Factor
    — Zero-product property.
    $2x - 11 = 0$ or $x + 3 = 0$
    $2x = 11$ or $x = -3$
    $x = \dfrac{11}{2}$

    $\left\{-3,\ \dfrac{11}{2}\right\}$

EXERCISES 8.3

1. $\sqrt{7}\sqrt{6} = \sqrt{42}$ — Since both radicands are positive we can multiply under the radical.

5. $\sqrt{7}\sqrt{7} = 7$ — This is the fundamental idea of radicals; $\sqrt{7}$ is one of two equal factors of 7.

9. $\sqrt{2}\sqrt{8} = \sqrt{2 \cdot 8}$ — Prime factor 8.
    $= \sqrt{2 \cdot 2^3}$
    $= \sqrt{2^4}$ — Add exponents. $2^2 \cdot 2^2 \cdot 2^4$.
    $= 2^2$
    $= 4$

13. $\sqrt{x}\sqrt{yz} = \sqrt{xyz}$ — Multiply radicands.

17. $\sqrt{5x}\sqrt{10y} = \sqrt{5x \cdot 2 \cdot 5y}$ — Prime factor 10.
    $= \sqrt{2 \cdot 5^2 xy}$ — Write in exponential form.
    $= 5\sqrt{2xy}$ — Simplify.

21. $\sqrt{11}\sqrt{10} = \sqrt{110}$ — Multiply radicands.

25. $(\sqrt{7} + \sqrt{3})(\sqrt{3})$
    $= \sqrt{7}\sqrt{3} + (\sqrt{3})^2$ — Distributive property.
    $= \sqrt{21} + 3$ — Simplify.

29. $\sqrt{10}(\sqrt{5} + \sqrt{2})$
    $= \sqrt{10}\sqrt{5} + \sqrt{10}\sqrt{2}$ — Distributive property.
    $= \sqrt{2 \cdot 5 \cdot 5} + \sqrt{2 \cdot 5 \cdot 2}$ — Prime factor 10.

Exercises 8.3

$= \sqrt{2 \cdot 5^2} + \sqrt{2^2 \cdot 5}$ Write the exponential form.
$= 5\sqrt{2} + 2\sqrt{5}$ Simplify.

33. $\sqrt{x^3}\sqrt{x^4} = \sqrt{x^2 \cdot x \cdot x^4}$ $x^3 = x^2 \cdot x$.
$\phantom{33. \sqrt{x^3}\sqrt{x^4}} = x \cdot x^2\sqrt{x}$ Simplify.
$\phantom{33. \sqrt{x^3}\sqrt{x^4}} = x^3\sqrt{x}$

37. $\sqrt{8r}\sqrt{125r^3} = \sqrt{8r \cdot 125r^3}$
$\phantom{37. \sqrt{8r}\sqrt{125r^3}} = \sqrt{2^3 \cdot 5^3 r^4}$ Prime factor both numbers.
$\phantom{37. \sqrt{8r}\sqrt{125r^3}} = 2 \cdot 5r^2 \sqrt{2 \cdot 5}$ Simplify.
$\phantom{37. \sqrt{8r}\sqrt{125r^3}} = 10r^2\sqrt{10}$ Multiply.

41. $\sqrt{2}(\sqrt{6} + \sqrt{8} + \sqrt{10})$
$ = \sqrt{2}\sqrt{6} + \sqrt{2}\sqrt{8} + \sqrt{2}\sqrt{10}$ Distributive property.
$ = \sqrt{2 \cdot 2 \cdot 3} + \sqrt{2 \cdot 2 \cdot 2 \cdot 2} + \sqrt{2 \cdot 2 \cdot 5}$ Prime factor 6, 8 and 10.
$ = \sqrt{2^2 \cdot 3} + \sqrt{2^4} + \sqrt{2^2 \cdot 5}$ Write in exponential form.
$ = 2\sqrt{3} + 2^2 + 2\sqrt{5}$ Simplify.
$ = 2\sqrt{3} + 4 + 2\sqrt{5}$

45. $3\sqrt{x}(\sqrt{x} + 3)$
$ = 3\sqrt{x}\sqrt{x} + 3\sqrt{x} \cdot 3$ Distributive property.
$ = 3x + 9\sqrt{x}$ Simplify.

49. $(\sqrt{5} - \sqrt{3})^2$
$ = (\sqrt{5})^2 - 2\sqrt{5}\sqrt{3} + (\sqrt{3})^2$ Recall $(a - b)^2 = a^2 - 2ab + b^2$.
$ = 5 - 2\sqrt{15} + 3$ Simplify.
$ = 8 - 2\sqrt{15}$ Combine like terms.

53. (a) $h = \dfrac{5}{2}\sqrt{3}$ Formula.

$ h = \dfrac{\sqrt{8}}{2}\sqrt{3}$ Replace s with $\sqrt{8}$.

$ h = \dfrac{\sqrt{2^3}}{2} \cdot \dfrac{\sqrt{3}}{1}$ Prime factor 8. $\sqrt{3} = \dfrac{\sqrt{3}}{1}$.

$ h = \dfrac{2\sqrt{2} \cdot \sqrt{3}}{2}$ Simplify and multiply.

$ h = \dfrac{2\sqrt{6}}{2}$ Multiply radicands.

$ h = \sqrt{6}$ Reduce.

$\sqrt{6}$ ft

(b) $\sqrt{6}$ ft ≈ 2.4 ft

57. (a) $A = wl$ Formula.
$A = (\sqrt{12})(\sqrt{15})$ Replace w with $\sqrt{12}$ and l with $\sqrt{15}$.

$A = \sqrt{12 \cdot 15}$ Multiply radicands.

$A = \sqrt{2^2 \cdot 3 \cdot 3 \cdot 5}$ Prime factor 12 and 15.
$A = \sqrt{2^2 \cdot 3^2 \cdot 5}$ Write in exponential form.

$A = 2 \cdot 3 \sqrt{5}$ Simplify.

$A = 6\sqrt{5}$

$6\sqrt{5}$ in^2

(b) $6\sqrt{5}$ in$^2 \approx 13.4$ in^2

61. $V = lwh$ Formula.
$V = \sqrt{8}\sqrt{2}\sqrt{5}$ Replace l with $\sqrt{8}$, w with $\sqrt{2}$ and h with $\sqrt{5}$.

$V = \sqrt{8 \cdot 2 \cdot 5}$ Multiply radicands.

$V = \sqrt{2^3 \cdot 2 \cdot 5}$ Prime factor 8.

$V = \sqrt{2^4 \cdot 5}$ $2^3 \cdot 2 = 2^4$.

$V = 2^2\sqrt{5}$ Simplify.

$V = 4\sqrt{5}$

$4\sqrt{5}$ cm$^3 \approx 8.9$ cm^3

65. $(\sqrt{60} - \sqrt{42})(\sqrt{60} + \sqrt{42})$
$= (\sqrt{60})^2 - (\sqrt{42})^2$ When you multiply conjugates the result is the difference of two squares.
$= 60 - 42$ Simplify.
$= 18$

73. $\dfrac{a-b}{a+b} - \dfrac{a+b}{a-b}$

$= \dfrac{a-b}{a+b} \cdot \dfrac{a-b}{a-b} - \dfrac{a+b}{a-b} \cdot \dfrac{a+b}{a+b}$ Write each fraction with the common denominator $(a+b)(a-b)$.

$= \dfrac{a^2 - 2ab + b^2}{(a+b)(a-b)} - \dfrac{a^2 + 2ab + b^2}{(a+b)(a-b)}$

$= \dfrac{a^2 - 2ab + b^2 - (a^2 + 2ab + b^2)}{(a+b)(a-b)}$ Subtract.

$= \dfrac{a^2 - 2ab + b^2 - a^2 - 2ab - b^2}{(a+b)(a-b)}$ Change the subtraction to addition.

$= \dfrac{-4ab}{a^2 - b^2}$

$= -\dfrac{4ab}{a^2 - b^2}$ $\dfrac{-x}{y} = -\dfrac{x}{y}$.

Exercises 8.4

77. $\dfrac{2m}{m-1} - \dfrac{m+4}{m-3}$

$= \dfrac{2m}{m-1} \cdot \dfrac{m-3}{m-3} - \dfrac{m+4}{m-3} \cdot \dfrac{m-1}{m-1}$ Write each fraction with the common denominator $(m-3)(m-1)$.

$= \dfrac{2m^2 - 6m}{(m-1)(m-3)} - \dfrac{m^2 + 3m - 4}{(m-1)(m-3)}$

$= \dfrac{2m^2 - 6m - (m^2 + 3m - 4)}{(m-1)(m-3)}$ Subtract.

$= \dfrac{2m^2 - 6m - m^2 - 3m + 4}{(m-1)(m-3)}$ Change the subtraction to addition.

$= \dfrac{m^2 - 9m + 4}{m^2 - 4m + 3}$

EXERCISES 8.4

1. $\dfrac{1}{\sqrt{3}} = \dfrac{1}{\sqrt{3}} \cdot \dfrac{\sqrt{3}}{\sqrt{3}}$ Multiply both the numerator and the denominator by $\sqrt{3}$.

$= \dfrac{\sqrt{3}}{3}$

5. $\dfrac{3}{\sqrt{11}} = \dfrac{3}{\sqrt{11}} \cdot \dfrac{\sqrt{11}}{\sqrt{11}}$ Multiply both the numerator and the denominator by $\sqrt{11}$.

$= \dfrac{3\sqrt{11}}{11}$

9. $\dfrac{\sqrt{6}}{\sqrt{9}} = \sqrt{\dfrac{6}{9}}$ Write under one radical and reduce.

$= \sqrt{\dfrac{2}{3}}$

$= \dfrac{\sqrt{2}}{\sqrt{3}} \cdot \dfrac{\sqrt{3}}{\sqrt{3}}$ Rationalize.

$= \dfrac{\sqrt{6}}{3}$

13. $\dfrac{3}{\sqrt{6}} = \dfrac{3}{\sqrt{6}} \cdot \dfrac{\sqrt{6}}{\sqrt{6}}$ Multiply both the numerator and the denominator by $\sqrt{6}$.

$= \dfrac{3\sqrt{6}}{6}$

$= \dfrac{\sqrt{6}}{2}$ Reduce.

17. $\dfrac{\sqrt{6}}{\sqrt{8}} = \sqrt{\dfrac{6}{8}}$ Write under one radical and reduce.

$= \sqrt{\dfrac{3}{4}}$

$= \dfrac{\sqrt{3}}{\sqrt{4}}$

$= \dfrac{\sqrt{3}}{2}$ Simplify.

21. $\dfrac{\sqrt{12}}{\sqrt{18}} = \sqrt{\dfrac{12}{18}}$ Write under one radical and reduce.

$= \sqrt{\dfrac{2}{3}}$

$= \dfrac{\sqrt{2}}{\sqrt{3}} \cdot \dfrac{\sqrt{3}}{\sqrt{3}}$ Rationalize.

$= \dfrac{\sqrt{6}}{3}$

25. $\sqrt{24} \div \sqrt{30} = \dfrac{\sqrt{24}}{\sqrt{30}}$ Write as a fraction.

$= \sqrt{\dfrac{24}{30}}$ Write under one radical and reduce.

$= \sqrt{\dfrac{4}{5}}$

$= \dfrac{\sqrt{4}}{\sqrt{5}} \cdot \dfrac{\sqrt{5}}{\sqrt{5}}$ Rationalize.

$= \dfrac{2\sqrt{5}}{5}$ Simplify.

29. $\dfrac{\sqrt{5}}{\sqrt{32}} = \dfrac{\sqrt{5}}{4\sqrt{2}}$ Simplify $\sqrt{32}$.

$= \dfrac{\sqrt{5}}{4\sqrt{2}} \cdot \dfrac{\sqrt{2}}{\sqrt{2}}$ Multiply both the numerator and the denominator by $\sqrt{2}$.

$= \dfrac{\sqrt{10}}{4 \cdot 2}$

$= \dfrac{\sqrt{10}}{8}$

Exercises 8.4

33. $\sqrt{18} \div \sqrt{81} = \dfrac{\sqrt{18}}{\sqrt{81}}$ Write as a fraction.

$\phantom{\sqrt{18} \div \sqrt{81}} = \dfrac{3\sqrt{2}}{9}$ Simplify $\sqrt{18}$ and $\sqrt{81}$.

$\phantom{\sqrt{18} \div \sqrt{81}} = \dfrac{\sqrt{2}}{3}$ Reduce.

37. $\dfrac{3\sqrt{6}}{7\sqrt{8}} = \dfrac{3}{7} \cdot \sqrt{\dfrac{6}{8}}$ Write under one radical and reduce.

$\phantom{\dfrac{3\sqrt{6}}{7\sqrt{8}}} = \dfrac{3}{7} \cdot \sqrt{\dfrac{3}{4}}$

$\phantom{\dfrac{3\sqrt{6}}{7\sqrt{8}}} = \dfrac{3}{7} \cdot \dfrac{\sqrt{3}}{2}$ Simplify.

$\phantom{\dfrac{3\sqrt{6}}{7\sqrt{8}}} = \dfrac{3\sqrt{3}}{14}$

41. $\dfrac{\sqrt{5}}{\sqrt{196}} = \dfrac{\sqrt{5}}{14}$ Simplify $\sqrt{196}$.

45. $\dfrac{2\sqrt{3}}{3\sqrt{40}} = \dfrac{2\sqrt{3}}{3 \cdot 2\sqrt{10}}$ Simplify $\sqrt{40}$.

$\phantom{\dfrac{2\sqrt{3}}{3\sqrt{40}}} = \dfrac{\sqrt{3}}{3\sqrt{10}}$ Reduce.

$\phantom{\dfrac{2\sqrt{3}}{3\sqrt{40}}} = \dfrac{\sqrt{3}}{3\sqrt{10}} \cdot \dfrac{\sqrt{10}}{\sqrt{10}}$ Multiply both the numerator and the denominator by $\sqrt{10}$.

$\phantom{\dfrac{2\sqrt{3}}{3\sqrt{40}}} = \dfrac{\sqrt{30}}{3 \cdot 10}$

$\phantom{\dfrac{2\sqrt{3}}{3\sqrt{40}}} = \dfrac{\sqrt{30}}{30}$

49. $\dfrac{\sqrt{32}}{3\sqrt{96}} = \dfrac{1}{3} \cdot \sqrt{\dfrac{32}{96}}$ Write as one radical and reduce.

$\phantom{\dfrac{\sqrt{32}}{3\sqrt{96}}} = \dfrac{1}{3} \cdot \sqrt{\dfrac{1}{3}}$

$\phantom{\dfrac{\sqrt{32}}{3\sqrt{96}}} = \dfrac{1}{3} \cdot \dfrac{1}{\sqrt{3}} \cdot \dfrac{\sqrt{3}}{\sqrt{3}}$ Multiply both the numerator and the denominator by $\sqrt{3}$.

$$= \frac{\sqrt{3}}{3 \cdot 3}$$ Simplify.

$$= \frac{\sqrt{3}}{9}$$

53. $\dfrac{5\sqrt{40}}{\sqrt{80}} = 5\sqrt{\dfrac{40}{80}}$ Write as one radical and reduce.

$$= 5\sqrt{\frac{1}{2}}$$

$$= 5\frac{\sqrt{1}}{\sqrt{2}} \cdot \frac{\sqrt{2}}{\sqrt{2}}$$ Multiply both the numerator and denominator by $\sqrt{2}$.

$$= 5\frac{\sqrt{2}}{2}$$

$$= \frac{5\sqrt{2}}{2}$$

57. $T = 2\pi \dfrac{\sqrt{L}}{\sqrt{32}}$ Formula.

$T = 2\pi \dfrac{\sqrt{2}}{\sqrt{32}}$ Replace L with 2.

$T = 2\pi \sqrt{\dfrac{2}{32}}$ Write as one radical and reduce.

$T = 2\pi \sqrt{\dfrac{1}{16}}$

$t = 2\pi \cdot \dfrac{1}{4}$ $\sqrt{\dfrac{1}{16}} = \dfrac{1}{4}$.

$T = \dfrac{\pi}{2}$ Reduce.

$\dfrac{\pi}{2}$ sec ≈ 1.57 sec

61. $s = \dfrac{\sqrt{M_1 M_2}}{\sqrt{F}}$ Formula.

$s = \dfrac{\sqrt{12 \cdot 4}}{\sqrt{20}}$ Replace M_1 with 12, M_2 with 4 and F with 20.

$s = \sqrt{\dfrac{12 \cdot 4}{20}}$ Write as one radical and reduce.

Exercises 8.4

$s = \sqrt{\dfrac{12}{5}}$

$s = \dfrac{\sqrt{12}}{\sqrt{5}}$ Simplify $\sqrt{12}$.

$s = \dfrac{2\sqrt{3}}{\sqrt{5}}$ Multiply both the numerator and the denominator by $\sqrt{5}$.

$s = \dfrac{2\sqrt{3}}{\sqrt{5}} \cdot \dfrac{\sqrt{5}}{\sqrt{5}}$

$s = \dfrac{2\sqrt{15}}{5}$ Simplify.

$\dfrac{2\sqrt{15}}{5} \approx 1.5$

65. $\dfrac{\sqrt{5} - \sqrt{2}}{\sqrt{5} + \sqrt{2}}$

$= \dfrac{\sqrt{5} - \sqrt{2}}{\sqrt{5} + \sqrt{2}} \cdot \dfrac{\sqrt{5} - \sqrt{2}}{\sqrt{5} - \sqrt{2}}$ Multiply both the numerator and the denominator by the conjugate of the denominator.

$= \dfrac{(\sqrt{5})^2 - 2\sqrt{5}\sqrt{2} + (\sqrt{2})^2}{(\sqrt{5})^2 - (\sqrt{2})^2}$

$= \dfrac{5 - 2\sqrt{10} + 2}{5 - 2}$ Simplify.

$= \dfrac{7 - 2\sqrt{10}}{3}$

69. $\dfrac{2x}{a + b} - \dfrac{3}{a + b}$

$= \dfrac{2x - 3}{a + b}$ Subtract the numerators.

73. $\dfrac{x + 1}{x^2 + 2x - 3} + \dfrac{x + 4}{x + 3} - \dfrac{5}{x - 1}$

$= \dfrac{x + 1}{(x + 3)(x - 1)} + \dfrac{x + 4}{x + 3} - \dfrac{5}{x - 1}$ Factor.

$= \dfrac{x + 1}{(x + 3)(x - 1)} + \dfrac{x + 4}{x + 3} \cdot \dfrac{x - 1}{x - 1}$ Write each fraction with the common denominator $(x + 3)(x - 1)$.

$- \dfrac{5}{x - 1} \cdot \dfrac{x + 3}{x + 3}$

$$= \frac{x+1}{(x+3)(x-1)} + \frac{x^2+3x-4}{(x+3)(x-1)} - \frac{5x+15}{(x+3)(x-1)}$$

$$= \frac{x+1+x^2+3x-4-(5x+15)}{(x+3)(x-1)} \qquad \text{Add and subtract.}$$

$$= \frac{x^2+4x-3-5x-15}{(x+3)(x-1)} \qquad \text{Combine like terms and change the subtraction to addition.}$$

$$= \frac{x^2-x-18}{(x+3)(x-1)}$$

77. $\dfrac{\dfrac{1}{2} - \dfrac{1}{b}}{\dfrac{1}{3} - \dfrac{1}{b}}$

$$= \frac{\dfrac{1}{2} - \dfrac{1}{b}}{\dfrac{1}{3} - \dfrac{1}{b}} \cdot \frac{6b}{6b} \qquad \text{Multiply both the numerator and the denominator by the LCM } 6b.$$

$$= \frac{3b-6}{2b-6} \qquad \text{Simplify.}$$

$$\text{or } \frac{3(b-2)}{2(b-3)}$$

EXERCISES 8.5

1. $3\sqrt{5} + 6\sqrt{5} = (3+6)\sqrt{5}$ Distributive property.
   $= 9\sqrt{5}$

5. $9\sqrt{2} - 7\sqrt{2} = (9-7)\sqrt{2}$ Distributive property.
   $= 2\sqrt{2}$

9. $6\sqrt{3} + \sqrt{3} = (6+1)\sqrt{3}$ Distributive property.
   $= 7\sqrt{3}$

13. $\sqrt{2} + \sqrt{2} = (1+1)\sqrt{2}$ Distributive property.
    $= 2\sqrt{2}$

17. $3\sqrt{2} + \sqrt{8} = 3\sqrt{2} + 2\sqrt{2}$ Simplify $\sqrt{8}$.
    $= (3+2)\sqrt{2}$ Distributive property.
    $= 5\sqrt{2}$

21. $\sqrt{45} + \sqrt{20} = 3\sqrt{5} + 2\sqrt{5}$ Simplify $\sqrt{45}$ and $\sqrt{20}$.
    $= (3+2)\sqrt{5}$ Distributive property.
    $= 5\sqrt{5}$

Exercises 8.5

25. $\sqrt{20} + \sqrt{45} - \sqrt{5}$
    $= 2\sqrt{5} + 3\sqrt{5} - \sqrt{5}$ Simplify $\sqrt{20}$ and $\sqrt{45}$.
    $= (2 + 3 - 1)\sqrt{5}$ Distributive property.
    $= 4\sqrt{5}$

29. $\sqrt{18x} + \sqrt{32x} = 3\sqrt{2x} + 4\sqrt{2x}$ Simplify $\sqrt{18x}$ and $\sqrt{32x}$.
    $\phantom{\sqrt{18x} + \sqrt{32x}} = (3 + 4)\sqrt{2x}$ Distributive property.
    $\phantom{\sqrt{18x} + \sqrt{32x}} = 7\sqrt{2x}$

33. $-x^2\sqrt{75} + x^2\sqrt{243}$
    $= -x^2 \cdot 5\sqrt{3} + x^2 \cdot 9\sqrt{3}$ Simplify $\sqrt{75}$ and $\sqrt{243}$.
    $= (-5x^2 + 9x^2)\sqrt{3}$ Distributive property.
    $= 4x^2\sqrt{3}$

37. $5\sqrt{56} + 8\sqrt{224}$
    $= 5 \cdot 2\sqrt{14} + 8 \cdot 4\sqrt{14}$ Simplify $\sqrt{56}$ and $\sqrt{224}$.
    $= 10\sqrt{14} + 32\sqrt{14}$ Multiply.
    $= (10 + 32)\sqrt{14}$ Distributive property.
    $= 42\sqrt{14}$

41. $\sqrt{72} - \sqrt{75} + \sqrt{98} - \sqrt{27}$
    $= 6\sqrt{2} - 5\sqrt{3} + 7\sqrt{2} - 3\sqrt{3}$ Simplify each radical.
    $= 13\sqrt{2} - 8\sqrt{3}$ Combine like radicals.

45. $x\sqrt{28y} - x\sqrt{700y} + x\sqrt{63y}$
    $= x \cdot 2\sqrt{7y} - x \cdot 10\sqrt{7y} + x \cdot 3\sqrt{7y}$ Simplify the radicals.
    $= (2x - 10x + 3x)\sqrt{7y}$ Distributive property.
    $= -5x\sqrt{7y}$

49. $3\sqrt{45} - 8\sqrt{48} + 11\sqrt{320}$
    $= 3 \cdot 3\sqrt{5} - 8 \cdot 4\sqrt{3} + 11 \cdot 8\sqrt{5}$ Simplify the radicals.
    $= 9\sqrt{5} - 32\sqrt{3} + 88\sqrt{5}$
    $= 97\sqrt{5} - 32\sqrt{3}$ Combine like radicals.

53. $3\sqrt{\dfrac{3}{2}} + 4\sqrt{\dfrac{1}{6}} - 5\sqrt{\dfrac{2}{3}}$

    $= 3\dfrac{\sqrt{3}}{\sqrt{2}} \cdot \dfrac{\sqrt{2}}{\sqrt{2}} + 4\dfrac{1}{\sqrt{6}} \cdot \dfrac{\sqrt{6}}{\sqrt{6}} - 5\dfrac{\sqrt{2}}{\sqrt{3}} \cdot \dfrac{\sqrt{3}}{\sqrt{3}}$ Rationalize the denominators.

    $= \dfrac{3\sqrt{6}}{2} + \dfrac{4\sqrt{6}}{6} - \dfrac{5\sqrt{6}}{3}$

    $= \dfrac{3\sqrt{6}}{2} \cdot \dfrac{3}{3} + \dfrac{4\sqrt{6}}{6} - \dfrac{5\sqrt{6}}{3} \cdot \dfrac{2}{2}$ Write each fraction with the common denominator 6.

    $= \dfrac{9\sqrt{6}}{6} + \dfrac{4\sqrt{6}}{6} - \dfrac{10\sqrt{6}}{6}$

$= \dfrac{9\sqrt{6} + 4\sqrt{6} - 10\sqrt{6}}{6}$ Add and subtract.

$= \dfrac{3\sqrt{6}}{6}$ Combine the radicals.

$= \dfrac{\sqrt{6}}{2}$ Reduce.

57. $\dfrac{1}{R} = \dfrac{1}{R_1} + \dfrac{1}{R_2}$ Formula.

$\dfrac{1}{R} = \dfrac{1}{\sqrt{\dfrac{3}{5}}} + \dfrac{1}{\sqrt{\dfrac{5}{3}}}$ Replace R_1 with $\sqrt{\dfrac{3}{5}}$ and R_2 with $\sqrt{\dfrac{5}{3}}$.

$\dfrac{1}{R} = \dfrac{1}{\dfrac{\sqrt{3}}{\sqrt{5}}} + \dfrac{1}{\dfrac{\sqrt{5}}{\sqrt{3}}}$

$\dfrac{1}{R} = \dfrac{\sqrt{5}}{\sqrt{3}} + \dfrac{\sqrt{3}}{\sqrt{5}}$ Multiply by the reciprocal of the denominator to simplify each complex fraction.

$\dfrac{1}{R} = \dfrac{\sqrt{5}}{\sqrt{3}} \cdot \dfrac{\sqrt{5}}{\sqrt{5}} + \dfrac{\sqrt{3}}{\sqrt{5}} \cdot \dfrac{\sqrt{3}}{\sqrt{3}}$ Write each fraction on the right side with the common denominator $\sqrt{3}\sqrt{5}$.

$\dfrac{1}{R} = \dfrac{5}{\sqrt{15}} + \dfrac{3}{\sqrt{15}}$

$\dfrac{1}{R} = \dfrac{8}{\sqrt{15}}$ Add.

$R = \dfrac{\sqrt{15}}{8}$ Write the reciprocal of each side.

61. $A = A_1 + A_2$ Add the areas of the two equilateral triangles.

$A = \dfrac{1}{4}s_1^2 \sqrt{3} + \dfrac{1}{4}s_2^2 \sqrt{3}$ Formula for area.

$A = \dfrac{1}{4}(25)^2 \sqrt{3} + \dfrac{1}{4}(20)^2 \sqrt{3}$ Replace s_1 with 25 and s_2 with 20.

$A = \dfrac{625\sqrt{3}}{4} + \dfrac{400\sqrt{3}}{4}$ Evaluate exponents.

$A = \dfrac{1025\sqrt{3}}{4}$ Add.

$\dfrac{1025\sqrt{3}}{4}$ yd^2 ≈ 443.8 yd^2

Exercises 8.6

65. $\sqrt{56x^2y} - x\sqrt{126y}$
    $= 2x\sqrt{14y} - 3x\sqrt{14y}$ Simplify $\sqrt{56x^2y}$ and $\sqrt{126y}$.
    $= (2x - 3x)\sqrt{14y}$ Distributive property.
    $= -x\sqrt{14y}$

69. $\dfrac{12}{5} = \dfrac{y}{35}$

    $5y = 420$ Cross multiply.
    $y = 84$ Solve for y.

73. $\dfrac{5}{x} = \dfrac{7.5}{9}$

    $7.5x = 45$ Cross multiply.
    $x = 6$ Solve for x.

77. $\dfrac{3}{5} = \dfrac{x}{5000 - x}$ One's share is x and the others's share is $5000 - x$.

    $5x = 3(5000 - x)$ Cross multiply.
    $5x = 15000 - 3x$
    $8x = 15000$
    $x = 1875$
    $5000 - x = 5000 - 1875 = 3125$

    $1875 and $3125

EXERCISES 8.6

1. $\sqrt{x} = 3$
   $(\sqrt{x})^2 = 3^2$ Square both sides.
   $x = 9$

   Check:
   $\sqrt{9} = 3$ is true.   Substitute 9 for x into the original equation.

   $\{9\}$

5. $\sqrt{2w} = 5$
   $(\sqrt{2w})^2 = 5^2$ Square both sides.
   $2w = 25$
   $w = \dfrac{25}{2}$

   Check:
   $\sqrt{2 \cdot \dfrac{25}{2}} = 5$ Substitute $\dfrac{25}{2}$ for w into the original equation.

   $\sqrt{25} = 5$ is true.

   $\left\{\dfrac{25}{2}\right\}$

9. $\sqrt{x + 3} = 5$
   $(\sqrt{x + 3})^2 = 5^2$ Square both sides.
   $x + 3 = 25$
   $x = 22$ Simplify.

   Check:

   $\sqrt{22 + 3} = 5$ Substitute 22 for x into the original equation.

   $\sqrt{25} = 5$ is true.

   $\{22\}$

13. $\sqrt{3w} = -9$
    $(\sqrt{3w})^2 = (-9)^2$ Square both sides.
    $3w = 81$
    $w = 27$

    Check:

    $\sqrt{3 \cdot 27} = -9$ Substitute 27 for w into the original equation.

    $\sqrt{81} = -9$ is false.   27 is an extraneous root.

    $\emptyset$

17. $\sqrt{3x + 1} = 5$
    $(\sqrt{3x + 1})^2 = 5^2$
    $3x + 1 = 25$ Square both sides.
    $3x = 24$
    $x = 8$

    Check:

    $\sqrt{3 \cdot 8 + 1} = 5$ Substitute 8 for x into the original equation.

    $\sqrt{25} = 5$ is true.

    $\{8\}$

21. $\sqrt{m - 7} + 8 = 5$
    $\sqrt{m - 7} = -3$ Subtract 8 from both sides.
    $(\sqrt{m - 7})^2 = (-3)^2$ Square both sides.
    $m - 7 = 9$
    $m = 16$

    Check:

    $\sqrt{16 - 7} + 8 = 5$ Substitute 16 for m into the original equation.
    $\sqrt{9} + 8 = 5$
    $3 + 8 = 5$
    $11 = 5$ False, so $m = 16$ is an extraneous root.

    $\emptyset$

Exercises 8.6

25. $5 - \sqrt{x+1} = -9$
　　　$-\sqrt{x+1} = -14$ 　　　　Subtract 5 from both sides.
　　$(-\sqrt{x+1})^2 = (-14)^2$ 　　Square both sides.
　　　　$x + 1 = 196$
　　　　　　$x = 195$

Check:

$5 - \sqrt{195 + 1} = -9$ 　　　Substitute 195 for x into the original equation.
$5 - \sqrt{196} = -9$
$5 - 14 = -9$
$-9 = -9$ is true.

$\{195\}$

29. $\sqrt{t} + 3\sqrt{2} = 5\sqrt{2}$
　　　　$\sqrt{t} = 2\sqrt{2}$ 　　　Subtract $3\sqrt{2}$ from both sides.
　　　$(\sqrt{t})^2 = (2\sqrt{2})^2$ 　　Square both sides.
　　　　　$t = 8$

Check:

$\sqrt{8} + 3\sqrt{2} = 5\sqrt{2}$ 　　Substitute 8 for t into the original equation.
$2\sqrt{2} + 3\sqrt{2} = 5\sqrt{2}$
$5\sqrt{2} = 5\sqrt{2}$ is true.

$\{8\}$

33. $7 - \sqrt{5x} = -1$
　　　$-\sqrt{5x} = -8$ 　　　Subtract 7 from both sides.
　　$(-\sqrt{5x})^2 = (-8)^2$ 　　Square both sides.
　　　　$5x = 64$

　　　　$x = \dfrac{64}{5}$

Check:

$7 - \sqrt{5 \cdot \dfrac{64}{5}} = -1$ 　　Substitute $\dfrac{64}{5}$ for x into the original equation.

$7 - \sqrt{64} = -1$
$7 - 8 = -1$
$-1 = -1$ is true.

$\left\{\dfrac{64}{5}\right\}$

37. $2\sqrt{x} + 3\sqrt{x} + 6 = 8$
　　　　$5\sqrt{x} + 6 = 8$ 　　Combine like radicals.
　　　　　$5\sqrt{x} = 2$ 　　　Subtract 6 from both sides.
　　　　$(5\sqrt{x})^2 = (2)^2$ 　　Square both sides.
　　　　　$25x = 4$

　　　　　　$x = \dfrac{4}{25}$

Check:

$$2\sqrt{\frac{4}{25}} + 3\sqrt{\frac{4}{25}} + 6 = 8$$

$$2\left(\frac{2}{5}\right) + 3\left(\frac{2}{5}\right) + 6 = 8$$

$$\frac{4}{5} + \frac{6}{5} + 6 = 8$$

$$8 = 8 \text{ is true.}$$

$\left\{\dfrac{4}{25}\right\}$

Substitute $\dfrac{4}{25}$ for x into the original equation.

41. $4\sqrt{y} - \sqrt{y} + 8 = 15$
$3\sqrt{y} + 8 = 15$ Combine like radicals.
$3\sqrt{y} = 7$ Subtract 8 from both sides.
$(3\sqrt{y})^2 = (7)^2$ Square both sides.
$9y = 49$
$y = \dfrac{49}{9}$

Check:

$$4\sqrt{\frac{49}{9}} - \sqrt{\frac{49}{9}} + 8 = 15$$

$$4\left(\frac{7}{3}\right) - \frac{7}{3} + 8 = 15$$

$$\frac{28}{3} - \frac{7}{3} + 8 = 15$$

$$15 = 15 \text{ is true.}$$

$\left\{\dfrac{49}{9}\right\}$

Substitute $\dfrac{49}{9}$ for y into the original equation.

45. $2\sqrt{t} - 7\sqrt{t} + 8 = 3\sqrt{t} - 8$
$-5\sqrt{t} + 8 = 3\sqrt{t} - 8$ Combine like radicals on the left side.
$8 = 8\sqrt{t} - 8$ Add $5\sqrt{t}$ to both sides.
$16 = 8\sqrt{t}$ Add 8 to both sides.
$2 = \sqrt{t}$ Divide both sides by 8.
$(2)^2 = (\sqrt{t})^2$ Square both sides.
$4 = t$

Check:

$2\sqrt{4} - 7\sqrt{4} + 8 = 3\sqrt{4} - 8$
$4 - 14 + 8 = 6 - 8$
$-2 = -2$ is true.

$\{4\}$

Substitute 4 for t into the original equation.

Exercises 8.6

49. $-6\sqrt{5a - 1} = -18$
$\sqrt{5a - 1} = 3$ Divide both sides by -6.
$(\sqrt{5a - 1})^2 = 3^2$ Square both sides.
$5a - 1 = 9$
$5a = 10$
$a = 2$

Check:

$-6\sqrt{5 \cdot 2 - 1} = -18$ Substitute 2 for a into the original equation.
$-6\sqrt{9} = -18$
$-18 = -18$ is true.

$\{2\}$

53. $\sqrt{x + 3} = \sqrt{2x - 1}$
$(\sqrt{x + 3})^2 = (\sqrt{2x - 1})^2$ Square both sides.
$x + 3 = 2x - 1$
$3 = x - 1$
$4 = x$

Check:

$\sqrt{4 + 3} = \sqrt{2 \cdot 4 - 1}$ Substitute 4 for x into the original equation.
$\sqrt{7} = \sqrt{7}$ is true.

$\{4\}$

57. $A = wl$ Formula.
$\sqrt{14} = \sqrt{y - 2}\sqrt{y + 3}$ Replace A with $\sqrt{14}$, w with $\sqrt{y - 2}$ and l with $\sqrt{y + 3}$.
$(\sqrt{14})^2 = (\sqrt{y - 2}\sqrt{y + 3})^2$ Square both sides.
$14 = (\sqrt{y - 2})^2(\sqrt{y + 3})^2$ Square of a product.
$14 = (y - 2)(y + 3)$
$14 = y^2 + y - 6$ Multiply using FOIL.
$0 = y^2 + y - 20$
$0 = (y + 5)(y - 4)$ Factor.
     Zero-product property.
$y + 5 = 0$ or $y - 4 = 0$
$y = -5$ or $y = 4$

Since the radicands must be non-negative, $y = -5$ is an extraneous root.
Check: $y = 4$

$\sqrt{14} = \sqrt{4 - 2}\sqrt{4 + 3}$
$\sqrt{14} = \sqrt{2}\sqrt{7}$
$\sqrt{14} = \sqrt{14}$ is true.

width: $\sqrt{y - 2} = \sqrt{4 - 2} = \sqrt{2}$ in.
length: $\sqrt{y + 3} = \sqrt{4 + 3} = \sqrt{7}$ in.

61. $\text{bore} = \sqrt{\dfrac{1}{0.7854} \cdot \dfrac{\text{displacement}}{\text{\# of cylinders}} \cdot \dfrac{1}{\text{stroke}}}$ Formula.

$3.5 = \sqrt{\dfrac{1}{0.7854} \cdot \dfrac{180}{4} \cdot \dfrac{1}{s}}$ Replace bore with 3.5, displacement with 180 and # of cylinders with 4.

$(3.5)^2 = \left(\sqrt{\dfrac{180}{3.1416s}}\right)^2$ Square both sides.

$12.25 = \dfrac{180}{3.1416s}$

$38.4846s = 180$ Cross multiply.
$s = 4.68$

4.68 in.

65. $\sqrt{12 - \sqrt{x}} = 3$
$\left(\sqrt{12 - \sqrt{x}}\right)^2 = (3)^2$ Square both sides.
$12 - \sqrt{x} = 9$
$-\sqrt{x} = -3$ Subtract 12 from both sides.
$(-\sqrt{x})^2 = (-3)^2$ Square both sides.
$x = 9$

Check:

$\sqrt{12 - \sqrt{9}} = 3$ Substitute 9 for x into the original equation.
$\sqrt{12 - 3} = 3$
$\sqrt{9} = 3$ is true.

$\{9\}$

69. $\begin{cases} x + y = 215 \\ x - y = 21 \end{cases}$

$2x = 236$ Add the two equations.
$x = 118$ Solve for x.

$118 + y = 215$ Replace x with 118 in Equation (1).
$y = 97$ Solve for y.

$\{(118, 97)\}$

73. $m = kn$ Equation of direct variation.
$20 = k(45)$ Replace m with 20 and n with 45.

$\dfrac{4}{9} = k$ Solve for k.

$m = \dfrac{4}{9}n$ Equation of variation with $\dfrac{4}{9}$ substituted for k.

$m = \dfrac{4}{9}(27)$ Replace n with 27.

$m = 12$ Solve for m.

77. salary: x
    sales: y

$x = ky$ Equation of direct variation.
$512 = k(6800)$ Replace x with 512 and y with 6800.

$\dfrac{32}{425} = k$ Solve for k.

$x = \dfrac{32}{425}y$ Equation of variation with $\dfrac{32}{425}$ substituted for k.

$x = \dfrac{32}{425}(21500)$ Replace y with 21500.

$x = 1619$ Solve for x.

$1619

CHAPTER 8 CONCEPT REVIEW

1. False; $\sqrt{25} = 5$ since the radical sign means the non-negative root.

2. True

3. True

4. True

5. False; The sum of two radicals is not the radical of the sum.

6. True

7. True

8. False; To combine like radical expressions, combine the coefficients and multiply by the common radical.

9. True

10. True

11. False; The first step is to isolate one of the radicals.

12. True

13. False; $\sqrt{125} = 5\sqrt{5}$.

14. True

15. False; If $a = b$, then $a^2 = b^2$.

CHAPTER 8 TEST

1. $\sqrt{\dfrac{7}{18}} = \dfrac{\sqrt{7}}{\sqrt{18}}$

   $= \dfrac{\sqrt{7}}{3\sqrt{2}}$ Simplify $\sqrt{18}$.

   $= \dfrac{\sqrt{7}}{3\sqrt{2}} \cdot \dfrac{\sqrt{2}}{\sqrt{2}}$ Multiply both the numerator and the denominator by $\sqrt{2}$.

   $= \dfrac{\sqrt{14}}{6}$ Simplify.

2. $\sqrt{y - 7} = 12$
   $(\sqrt{y - 7})^2 = 12^2$ Square both sides.
   $y - 7 = 144$
   $y = 151$

   Check:
   $\sqrt{151 - 7} = 12$ Substitute 151 for y into the original equation.
   $\sqrt{144} = 12$ is true.

   $\{151\}$

3. $-\sqrt{196} = -(14)$
   $= -14$ $14^2 = 196$.

4. $\sqrt{117} = \sqrt{9 \cdot 13}$ The largest square factor of 117 is 9, so write 117 as $9 \cdot 13$.
   $= 3\sqrt{13}$

5. $\dfrac{5}{\sqrt{12}} = \dfrac{5}{2\sqrt{3}}$ Simplify $\sqrt{12}$.

   $= \dfrac{5}{2\sqrt{3}} \cdot \dfrac{\sqrt{3}}{\sqrt{3}}$ Multiply both the numerator and the denominator by $\sqrt{3}$.

   $= \dfrac{5\sqrt{3}}{6}$

6. $5 + 3\sqrt{2x + 26} = 23$
   $3\sqrt{2x + 26} = 18$ Subtract 5 from both sides.
   $\sqrt{2x + 26} = 6$ Divide both sides by 3.
   $(\sqrt{2x + 26})^2 = 6^2$ Square both sides.
   $2x + 26 = 36$
   $2x = 10$
   $x = 5$

Check:

$5 + 3\sqrt{2 \cdot 5 + 26} = 23$
$5 + 3\sqrt{36} = 23$
$\phantom{5 + 3\sqrt{36}}5 + 3(6) = 23$
$\phantom{5 + 3\sqrt{36} = }23 = 23$ is true.

$\{5\}$

Substitute 5 for x into the original equation.

7. $\sqrt{10}(\sqrt{15} - \sqrt{18})$
   $= \sqrt{10}\sqrt{15} - \sqrt{10}\sqrt{18}$
   $= \sqrt{10 \cdot 15} - \sqrt{10 \cdot 18}$
   $= \sqrt{2 \cdot 5 \cdot 3 \cdot 5} - \sqrt{2 \cdot 5 \cdot 2 \cdot 3 \cdot 3}$
   $= \sqrt{2 \cdot 3 \cdot 5^2} - \sqrt{2^2 \cdot 3^2 \cdot 5}$
   $= 5\sqrt{6} - 2 \cdot 3\sqrt{5}$
   $= 5\sqrt{6} - 6\sqrt{5}$

   Distributive property.
   Multiply radicands.
   Prime factor the numbers.

   Write in exponential form.
   Simplify.

8. $3\sqrt{27} - 2\sqrt{12} - \sqrt{3}$
   $= 3 \cdot 3\sqrt{3} - 2 \cdot 2\sqrt{3} - \sqrt{3}$
   $= 9\sqrt{3} - 4\sqrt{3} - \sqrt{3}$
   $= (9 - 4 - 1)\sqrt{3}$
   $= 4\sqrt{3}$

   Simplify $\sqrt{27}$ and $\sqrt{12}$.

   Distributive property.

9. $\sqrt{179} \approx 13.38$

   Calculator.

10. $\sqrt{72} - \sqrt{98} + \sqrt{242}$
    $= 6\sqrt{2} - 7\sqrt{2} + 11\sqrt{2}$
    $= (6 - 7 + 11)\sqrt{2}$
    $= 10\sqrt{2}$

    Simplify the radicals.
    Distributive property.

11. $\sqrt{98pq^2} = \sqrt{2 \cdot 49pq^2}$
    $\phantom{\sqrt{98pq^2}} = 7q\sqrt{2p}$

    49 is the largest square factor of 98.
    Simplify.

12. $2\sqrt{8} \cdot 5\sqrt{162}$
    $= 2 \cdot 2\sqrt{2} \cdot 5 \cdot 9\sqrt{2}$
    $= 180 \cdot (\sqrt{2})^2$
    $= 180 \cdot 2$
    $= 360$

    Simplify $\sqrt{8}$ and $\sqrt{162}$.
    Multiply.

13. $A = \dfrac{1}{2}bh$

    Area formula.

    $A = \dfrac{1}{2}(\sqrt{30})(\sqrt{10})$

    Replace b with $\sqrt{30}$ and h with $\sqrt{10}$.

    $A = \dfrac{1}{2}\sqrt{3}\sqrt{10}\sqrt{10}$

    $\sqrt{30} = \sqrt{3} \cdot \sqrt{10}$.

$A = \dfrac{1}{2} \cdot \sqrt{3} \cdot 10$ $\qquad\qquad (\sqrt{10})^2 = 10.$

$A = 5\sqrt{3}$

$5\sqrt{3}$ ft^2

14. $V_1 = \sqrt{(\text{watts})(\text{ohms})}$ Formula.
$ = \sqrt{(500)(40)}$ Replace watts with 500 and ohms with 40.
$ = \sqrt{2^2 \cdot 5^3 \cdot 2^3 \cdot 5}$ Prime factor the numbers.
$ = \sqrt{2^5 \cdot 5^4}$
$ = 2^2 \cdot 5^2 \sqrt{2}$ Simplify.
$ = 100\sqrt{2}$

$V_2 = \sqrt{(\text{watts})(\text{ohms})}$
$ = \sqrt{(200)(25)}$ Replace watts with 200 and ohms with 25.
$ = \sqrt{2^3 \cdot 5^2 \cdot 5^2}$ Prime factor the numbers.
$ = \sqrt{2^3 \cdot 5^4}$
$ = 2 \cdot 5^2 \sqrt{2}$ Simplify.
$ = 50\sqrt{2}$

Total Voltage $= V_1 + V_2$
$\phantom{\text{Total Voltage}} = 100\sqrt{2} + 50\sqrt{2}$ Replace V_1 with $100\sqrt{2}$ and V_2 with $50\sqrt{2}$.
$\phantom{\text{Total Voltage}} = 150\sqrt{2}$ volts

CHAPTER 9

EXERCISES 9.1

1. $x(x - 3) = 0$
   $x = 0$ or $x - 3 = 0$
   $\phantom{x = 0 \text{ or }} x = 3$
   $\{0, 3\}$ Zero-product property.

5. $6x^2 + 18x = 0$
   $6x(x + 3) = 0$ Factor the left side.

   $6x = 0$ or $x + 3 = 0$ Zero-product property.
   $x = 0$ or $ x = -3$
   $\{-3, 0\}$

9. $x^2 = 4$
   $x = \pm\sqrt{4}$ Square root property: set x equal to the two square roots of 4.
   $x = \pm 2$
   $\{\pm 2\}$

13. $b^2 = -9$
    No real solution   If b^2 is equal to a negative number, the equation has no real solution.

17. $3(x^2 + 8) = 24 - 15x$
    $3x^2 + 24 = 24 - 15x$ Multiply to clear the parentheses.
    $3x^2 + 15x = 0$ Write in standard form.
    $3x(x + 5) = 0$ Factor.

    $3x = 0$ or $x + 5 = 0$ Zero-product property.
    $x = 0$ or $ x = -5$
    $\{-5, 0\}$

21. $(x + 3)^2 = 9 + 11x + 2x^2$
    $x^2 + 6x + 9 = 9 + 11x + 2x^2$ Multiply to clear the parentheses.
    $0 = x^2 + 5x$ Write in standard form.
    $0 = x(x + 5)$ Factor.

    $x = 0$ or $x + 5 = 0$ Zero-product property.
    $\phantom{x = 0 \text{ or } x + 5 =} x = -5$
    $\{-5, 0\}$

25. $2x^2 - 35 = x^2 - 9$ Isolate the term with the variable.
    $x^2 = 26$ Square root property: set x equal to the two square roots of 26.
    $x = \pm\sqrt{26}$
    $\{\pm\sqrt{26}\}$

29. $14 - 5x + 3x^2 = 24 - 5x + 2x^2$
    $x^2 = 10$ Isolate the term with the variable.
    $x = \pm\sqrt{10}$ Square root property: set x equal to the two square roots of 10.
    $\{\pm\sqrt{10}\}$

33. $\quad 4x^2 + 8x - 10 = x(x + 2) + 2(x - 5)$
$\quad\quad 4x^2 + 8x - 10 = x^2 + 2x + 2x - 10 \quad$ Multiply to clear the parentheses.
$\quad\quad 4x^2 + 8x - 10 = x^2 + 4x - 10 \quad$ Combine like terms.
$\quad\quad\quad\quad 3x^2 + 4x = 0 \quad$ Write in standard form.
$\quad\quad\quad\quad x(3x + 4) = 0 \quad$ Factor.

$x = 0 \quad$ or $\quad 3x + 4 = 0 \quad$ Zero-product property.
$\quad\quad\quad\quad\quad 3x = -4$
$\quad\quad\quad\quad\quad x = -\dfrac{4}{3}$

$\left\{-\dfrac{4}{3},\, 0\right\}$

37. $\quad \dfrac{21}{y - 3} = y - 7$

$(y - 3)\left(\dfrac{21}{y - 3}\right) = (y - 3)(y - 7) \quad$ Multiply both sides by $y - 3$ to clear the fractions.

$\quad\quad 21 = y^2 - 10y + 21 \quad$ Simplify.
$\quad\quad 0 = y^2 - 10y \quad$ Write in standard form.
$\quad\quad 0 = y(y - 10) \quad$ Factor.

$y = 0 \quad$ or $\quad y - 10 = 0 \quad$ Zero-product property.
$\quad\quad\quad\quad\quad y = 10$

$\{0,\, 10\}$

41. $V^2 = 23$
$\quad V = \pm\sqrt{23} \quad$ Square root property.
$\quad V = \pm 4.80 \quad$ Calculator.

$\{\pm 4.80\}$

45. $3x^2 - 17 = 20$
$\quad\quad 3x^2 = 37 \quad$ Isolate the variable.
$\quad\quad x^2 = \dfrac{37}{3}$

$\quad\quad x = \pm\sqrt{\dfrac{37}{3}} \quad$ Square root property.

$\quad\quad x = \pm 3.51 \quad$ Calculator.

$\{\pm 3.51\}$

49. $4x^2 = 100a^2, \quad a \geq 0$
$\quad x^2 = 25a^2$
$\quad x = \pm\sqrt{25a^2} \quad$ Isolate the variable.
$\quad x = \pm 5a \quad$ Square root property.

53. length of side: x
$\quad A = s^2 \quad$ Formula.
$\quad 2116 = x^2 \quad$ Replace A with 2116 and s with x.

Exercises 9.2

$\sqrt{2116} = x$ $x > 0$ since it represents a length.
$46 = x$

46 ft

57. $R = \dfrac{B^2 - D^2 - E^2}{2E}$ Formula.

$188 = \dfrac{B^2 - (23.1)^2 - (1.3)^2}{2(1.3)}$ Replace R with 188, D with 23.1 and E with 1.3.

$188 = \dfrac{B^2 - 533.61 - 1.69}{2.6}$

$488.8 = B^2 - 535.3$ Multiply both sides by 2.6. $-533.61 - 1.69 = -535.3$.
$1024.1 = B^2$ Isolate the variable.
$32.0 = B$ $B > 0$ since it is a distance.

32.0 ft

61. $(x - 5)^2 = 7$ Substitute y for $x - 5$.
         $y^2 = 7$
         $y = \pm\sqrt{7}$ Square root property.

$y = \sqrt{7}$ or $y = -\sqrt{7}$
$x - 5 = \sqrt{7}$ or $x - 5 = -\sqrt{7}$ Replace y with $x - 5$.
$x = 5 + \sqrt{7}$ or $x = 5 - \sqrt{7}$

$\{5 + \sqrt{7}, 5 - \sqrt{7}\}$

65. $\sqrt{169} = 13$ $13^2 = 169$.

69. $\sqrt{68} = \sqrt{4}\sqrt{17}$ 4 is the largest square factor of 68.
       $= 2\sqrt{17}$ Simplify.

73. $V = \sqrt{ER}$ Formula.
      $V = \sqrt{(1200)(30)}$ Replace E with 1200 and R with 30.
      $V = \sqrt{30 \cdot 4 \cdot 10 \cdot 30}$ $1200 = 30 \cdot 4 \cdot 10$.
      $V = \sqrt{4 \cdot (30)^2 \cdot 10}$
      $V = 2 \cdot 30\sqrt{10}$ Simplify.
      $V = 60\sqrt{10}$

$60\sqrt{10}$ volts

EXERCISES 9.2

1. $a^2 + b^2 = c^2$ Pythagorean formula.
     $3^2 + 4^2 = c^2$ Substitute $a = 3$ and $b = 4$.
     $9 + 16 = c^2$ Solve.
         $25 = c$
          $5 = c$ The negative root is not used in the solution of triangles.

5. $a^2 + b^2 = c^2$ Pythagorean formula.
$15^2 + b^2 = 39^2$ Substitute $a = 15$ and $c = 39$.
$225 + b^2 = 1521$ Solve.
$b^2 = 1296$
$b = 36$ The negative root is not used in the solution of triangles.

9. $a^2 + b^2 = c^2$ Pythagorean formula.
$a^2 + 30^2 = 34^2$ Substitute $b = 30$ and $c = 34$.
$a^2 + 900 = 1156$ Solve.
$a^2 = 256$
$a = 16$ The negative root is not used in the solution of triangles.

13. $a^2 + b^2 = c^2$ Pythagorean formula.
$(7.5)^2 + 18^2 = c^2$ Substitute $a = 7.5$ and $b = 18$.
$56.25 + 324 = c^2$ Solve.
$380.25 = c^2$
$19.5 = c$ The negative root is not used in the solution of triangles.

17. $a^2 + b^2 = c^2$ Pythagorean formula.
$a^2 + 8^2 = 17^2$ Substitute $b = 8$ and $c = 17$.
$a^2 + 64 = 289$ Solve.
$a^2 = 225$
$a = 15$ The negative root is not used in the solution of triangles.

21. $a^2 + b^2 = c^2$ Pythagorean formula.
$45^2 + 24^2 = c^2$ Substitute $a = 45$ and $b = 24$.
$2025 + 576 = c^2$ Solve.
$2601 = c^2$
$51 = c$ The negative root is not used in the solution of triangles.

25. $a^2 + b^2 = c^2$ Pythagorean formula.
$2^2 + b^2 = (4.25)^2$ Substitute $a = 2$ and $c = 4.25$.
$4 + b^2 = 18.0625$ Solve.
$b^2 = 14.0625$
$b = 3.75$ The negative root is not used in the solution of triangles.

29. $a^2 + b^2 = c^2$ Pythagorean formula.
$(13.5)^2 + 18^2 = c^2$ Substitute $a = 13.5$ and $b = 18$.
$182.25 + 324 = c^2$ Solve.
$506.25 = c^2$
$22.5 = c$ The negative root is not used in the solution of triangles.

33. $a^2 + b^2 = c^2$ Pythagorean formula.
$(1.5)^2 + b^2 = (2.5)^2$ Substitute $a = 1.5$ and $c = 2.5$.
$2.25 + b^2 = 6.25$ Solve.
$b^2 = 4$
$b = 2$ The negative root is not used in the solution of triangles.

37. $a^2 + b^2 = c^2$ Pythagorean formula.
$2^2 + 3^2 = c^2$ Substitute $a = 2$ and $b = 3$.
$4 + 9 = c^2$ Solve.

Exercises 9.2

$$13 = c^2$$
$$\sqrt{13} = c$$
$$c \approx 3.61$$

The negative root is not used in the solution of triangles.

41. $a^2 + b^2 = c^2$ Pythagorean formula.
    $2^2 + b^2 = 8^2$ Substitute $a = 2$ and $b = 3$.
    $4 + b^2 = 64$ Solve.
    $b^2 = 60$
    $b = \sqrt{60}$ The negative root is not used in the solution of triangles.
    $b = 2\sqrt{15}$ $60 = 4 \cdot 15$.
    $b \approx 7.75$

45. $a^2 + b^2 = c^2$ Pythagorean formula.
    $a^2 + 5^2 = (7.5)^2$ Substitute $b = 5$ and $c = 7.5$.
    $a^2 + 25 = 56.25$ Solve.
    $a^2 = 31.25$
    $a \approx 5.59$ The negative root is not used in the solution of triangles.

49. $a^2 + b^2 = c^2$ Pythagorean formula.
    $(0.5)^2 + (0.75)^2 = c^2$ Substitute $a = 0.5$ and $b = 0.75$.
    $0.25 + 0.5625 = c^2$ Solve.
    $0.8125 = c^2$
    $0.90 \approx c$ The negative root is not used in the solution of triangles.

53. $a^2 + b^2 = c^2$ Pythagorean formula.
    $a^2 + 8^2 = 11^2$ Substitute $b = 8$ and $c = 11$.
    $a^2 + 64 = 121$ Solve.
    $a^2 = 57$
    $a \approx 7.55$ The negative root is not used in the solution of triangles.

57. $a^2 + b^2 = c^2$ (Rise)^2 + (Run)^2 = (Length)^2.
    $5^2 + 13^2 = c^2$ Substitute $a = 5$ and $b = 13$.
    $25 + 169 = c^2$ Solve.
    $194 = c^2$
    $13.93 \approx c$ Length is positive.

    13.93 ft

61. $a^2 + b^2 = c^2$ Pythagorean formula.
    $10^2 + 22^2 = c^2$ Substitute $a = 10$ and $b = 22$.
    $100 + 484 = c^2$ Solve.
    $584 = c^2$
    $24.2 \approx c$ Length is positive.

    24.2 ft

65. $a^2 + b^2 = c^2$ Pythagorean formula.
    $90^2 + 90^2 = c^2$ Substitute $a = 90$ and $b = 90$.
    $8100 + 8100 = c^2$ Solve.
    $16200 = c^2$
    $127.3 \approx c$ Distance is positive.

    127.3 ft

69. $a^2 + b^2 = c^2$
    $\pm\sqrt{a^2 + b^2} = c$ Square root property.

73. $\sqrt{294a^5b^6c^3} = \sqrt{49a^4b^6c^2 \cdot 6ac}$ $294 = 49 \cdot 6$.
    $\phantom{\sqrt{294a^5b^6c^3}} = 7a^2b^3c\sqrt{6ac}$ Simplify.

77. $15\sqrt{40} \cdot 3\sqrt{15}$
    $= 15 \cdot 2\sqrt{10} \cdot 3\sqrt{15}$ Simplify $\sqrt{40}$.
    $= 90\sqrt{2 \cdot 5 \cdot 3 \cdot 5}$ $15 \cdot 2 \cdot 3 = 90$. Prime factor 10 and 15.
    $= 90 \cdot 5\sqrt{6}$ Simplify.
    $= 450\sqrt{6}$

81. $A = \dfrac{1}{2}bh$ Formula for area.

    $A = \dfrac{1}{2}\sqrt{30}\sqrt{70}$ Replace b with $\sqrt{30}$ and h with $\sqrt{70}$.

    $A = \dfrac{1}{2}\sqrt{30 \cdot 70}$

    $A = \dfrac{1}{2}\sqrt{3 \cdot 2 \cdot 5 \cdot 7 \cdot 2 \cdot 5}$ Prime factor 30 and 70.

    $A = \dfrac{1}{2}\sqrt{2^2 \cdot 3 \cdot 5^2 \cdot 7}$ Write in exponential form.

    $A = \dfrac{1}{2} \cdot 2 \cdot 5\sqrt{21}$ Simplify.

    $A = 5\sqrt{21}$

    $5\sqrt{21}$ ft ≈ 22.9 ft

EXERCISES 9.3

1. $(x + 3)^2 = 25$ There is a square binomial on the left.
    $x + 3 = \pm 5$ Square root property.

    $x + 3 = 5$ or $x + 3 = -5$ Solve each equation.
    $ x = 2$ or $ x = -8$
    $\{-8, 2\}$

5. $(x + 5)^2 = 20$ There is a square binomial on the left.
    $x + 5 = \pm 2\sqrt{5}$ Square root property.
    $x = -5 \pm 2\sqrt{5}$ Solve for x.
    $\{-5 \pm 2\sqrt{5}\}$

9. $x^2 + 10x - 24 = 0$
    $x^2 + 10x = 24$ Add 24 to both sides.
    $x^2 + 10x + 25 = 24 + 25$ Add 25 to both sides: $25 = \left[\dfrac{1}{2}(10)\right]^2 = (5)^2$.

    $(x + 5)^2 = 49$ Factor the left side and simplify the right side.
    $x + 5 = \pm 7$ Square root property.

Exercises 9.3

$$x + 5 = 7 \quad \text{or} \quad x + 5 = -7$$
$$x = 2 \quad \text{or} \quad x = -12 \qquad \text{Solve for } x.$$
$$\{-12, 2\}$$

13. $x^2 + 6x - 7 = 0$
$$x^2 + 6x = 7 \qquad \text{Add 7 to both sides.}$$
$$x^2 + 6x + 9 = 7 + 9 \qquad \text{Add 9 to both sides: } 9 = \left[\tfrac{1}{2}(6)\right]^2 = (3)^2.$$
$$(x + 3)^2 = 16 \qquad \text{Factor the left side and simplify the right side.}$$
$$x + 3 = \pm 4 \qquad \text{Square root property.}$$
$$x + 3 = 4 \quad \text{or} \quad x + 3 = -4$$
$$x = 1 \quad \text{or} \quad x = -7 \qquad \text{Solve for } x.$$
$$\{-7, 1\}$$

17. $\left(x - \tfrac{4}{3}\right)^2 = \tfrac{25}{9}$

$$x - \tfrac{4}{3} = \pm \tfrac{5}{3} \qquad \text{Square root property.}$$

$$x - \tfrac{4}{3} = \tfrac{5}{3} \quad \text{or} \quad x - \tfrac{4}{3} = -\tfrac{5}{3}$$

$$x = 3 \quad \text{or} \quad x = -\tfrac{1}{3} \qquad \text{Solve for } x.$$

$$\left\{-\tfrac{1}{3}, 3\right\}$$

21. $\left(y + \tfrac{11}{6}\right)^2 = \tfrac{11}{36}$

$$y + \tfrac{11}{6} = \pm \tfrac{\sqrt{11}}{6} \qquad \text{Square root property.}$$

$$y = -\tfrac{11}{6} \pm \tfrac{\sqrt{11}}{6} \qquad \text{Solve for } y.$$

$$y = \tfrac{-11 \pm \sqrt{11}}{6} \qquad \text{Simplify.}$$

$$\left\{\tfrac{-11 \pm \sqrt{11}}{6}\right\}$$

25. $x^2 + 5x + 5 = 0$
$$x^2 + 5x = -5 \qquad \text{Subtract 5 from both sides.}$$
$$x^2 + 5x + \tfrac{25}{4} = -5 + \tfrac{25}{4} \qquad \text{Add } \tfrac{25}{4} \text{ to both sides: } \tfrac{25}{4} = \left[\tfrac{1}{2}(5)\right]^2 = \left(\tfrac{5}{2}\right)^2.$$

$$\left(x + \frac{5}{2}\right)^2 = \frac{5}{4}$$ Factor the left side and simplify the right side.

$$x + \frac{5}{2} = \pm\frac{\sqrt{5}}{2}$$ Square root property.

$$x = -\frac{5}{2} \pm \frac{\sqrt{5}}{2}$$ Solve for x.

$$x = \frac{-5 \pm \sqrt{5}}{2}$$ Simplify.

$$\left\{\frac{-5 \pm \sqrt{5}}{2}\right\}$$

29. $2x^2 - 6x - 5 = 0$

$$x^2 - 3x - \frac{5}{2} = 0$$ Divide both sides by 2 so the coefficient of x^2 is 1.

$$x^2 - 3x = \frac{5}{2}$$ Add $\frac{5}{2}$ to both sides.

$$x^2 - 3x + \frac{9}{4} = \frac{5}{2} + \frac{9}{4}$$ Add $\frac{9}{4}$ to both sides: $\frac{9}{4} = \left[\frac{1}{2}(-3)\right]^2 = \left(-\frac{3}{2}\right)^2$.

$$\left(x - \frac{3}{2}\right)^2 = \frac{19}{4}$$ Factor the left side and simplify the right side.

$$x - \frac{3}{2} = \pm\frac{\sqrt{19}}{2}$$ Square root property.

$$x = \frac{3}{2} \pm \frac{\sqrt{19}}{2}$$ Solve for x.

$$x = \frac{3 \pm \sqrt{19}}{2}$$ Simplify.

$$\left\{\frac{3 \pm \sqrt{19}}{2}\right\}$$

33. $2x^2 + 15x - 27 = 0$

$$x^2 + \frac{15}{2}x - \frac{27}{2} = 0$$ Divide both sides by 2 so the coefficient of x^2 is 1.

$$x^2 + \frac{15}{2}x = \frac{27}{2}$$ Add $\frac{27}{2}$ to both sides.

$$x^2 + \frac{15}{2}x + \frac{225}{16} = \frac{27}{2} + \frac{225}{16}$$ Add $\frac{225}{16}$ to both sides: $\frac{225}{16} = \left[\frac{1}{2}\left(\frac{15}{2}\right)\right]^2 = \left(\frac{15}{4}\right)^2$.

$$\left(x + \frac{15}{4}\right)^2 = \frac{441}{16}$$ Factor the left side and simplify the right side.

$$x + \frac{15}{4} = \pm\frac{21}{4}$$ Square root property.

Exercises 9.3

$$x + \frac{15}{4} = \frac{21}{4} \quad \text{or} \quad x + \frac{15}{4} = -\frac{21}{4}$$

$$x = \frac{3}{2} \quad \text{or} \quad x = -9 \qquad \text{Solve for } x.$$

$$\left\{-9, \frac{3}{2}\right\}$$

37. $x^2 - 7x - 2 = 0$

$\qquad x^2 - 7x = 2$ Add 2 to both sides.

$\qquad x^2 - 7x + \dfrac{49}{4} = 2 + \dfrac{49}{4}$ Add $\dfrac{49}{4}$ to both sides: $\dfrac{49}{4} = \left[\dfrac{1}{2}(-7)\right]^2 = \left(-\dfrac{7}{2}\right)^2$.

$\qquad \left(x - \dfrac{7}{2}\right)^2 = \dfrac{57}{4}$ Factor the left side and simplify the right side.

$\qquad x - \dfrac{7}{2} = \pm \dfrac{\sqrt{57}}{2}$ Square root property.

$\qquad x = \dfrac{7}{2} \pm \dfrac{\sqrt{57}}{2}$ Solve for x.

$\qquad x = \dfrac{7 \pm \sqrt{57}}{2}$ Simplify.

$$\left\{\frac{7 \pm \sqrt{57}}{2}\right\}$$

41. $2x^2 - 19x + 35 = 0$

$\qquad x^2 - \dfrac{19}{2}x + \dfrac{35}{2} = 0$ Divide both sides by 2 so the coefficient of x^2 is 1.

$\qquad x^2 - \dfrac{19}{2}x = -\dfrac{35}{2}$ Subtract $\dfrac{35}{2}$ from both sides.

$\qquad x^2 - \dfrac{19}{2}x + \dfrac{361}{16} = -\dfrac{35}{2} + \dfrac{361}{16}$ Add $\dfrac{361}{16}$ to both sides: $\dfrac{361}{16} = \left[\dfrac{1}{2}\left(-\dfrac{19}{2}\right)\right]^2 = \left(-\dfrac{19}{4}\right)^2$.

$\qquad \left(x - \dfrac{19}{4}\right)^2 = \dfrac{81}{16}$ Factor the left side and simplify the right side.

$\qquad x - \dfrac{19}{4} = \pm \dfrac{9}{4}$ Square root property.

$$x - \frac{19}{4} = \frac{9}{4} \quad \text{or} \quad x - \frac{19}{4} = -\frac{9}{4}$$

$$x = 7 \quad \text{or} \quad x = \frac{5}{2} \qquad \text{Solve for } x.$$

$$\left\{\frac{5}{2}, 7\right\}$$

45. $5x^2 - 11x - 2 = 0$

$x^2 - \dfrac{11}{5}x - \dfrac{2}{5} = 0$ Divide by 5 so the coefficient of x^2 is 1.

$x^2 - \dfrac{11}{5}x = \dfrac{2}{5}$ Add $\dfrac{2}{5}$ to both sides.

$x^2 - \dfrac{11}{5}x + \dfrac{121}{100} = \dfrac{2}{5} + \dfrac{121}{100}$ Add $\dfrac{121}{100}$ to both sides: $\dfrac{121}{100} = \left[\dfrac{1}{2}\left(-\dfrac{11}{5}\right)\right]^2 = \left(-\dfrac{11}{10}\right)^2$.

$\left(x - \dfrac{11}{10}\right)^2 = \dfrac{161}{100}$ Factor the left side and simplify the right side.

$x - \dfrac{11}{10} = \pm\dfrac{\sqrt{161}}{10}$ Square root property.

$x = \dfrac{11}{10} \pm \dfrac{\sqrt{161}}{10}$ Solve for x.

$x = \dfrac{11 \pm \sqrt{161}}{10}$ Simplify.

$\left\{\dfrac{11 \pm \sqrt{161}}{10}\right\}$

49. $9x^2 - 4x + 1 = 0$

$x^2 - \dfrac{4}{9}x + \dfrac{1}{9} = 0$ Divide by 9 so the coefficient of x^2 is 1.

$x^2 - \dfrac{4}{9}x = -\dfrac{1}{9}$ Subtract $\dfrac{1}{9}$ from both sides.

$x^2 - \dfrac{4}{9}x + \dfrac{4}{81} = -\dfrac{1}{9} + \dfrac{4}{81}$ Add $\dfrac{4}{81}$ to both sides: $\dfrac{4}{81} = \left[\dfrac{1}{2}\left(-\dfrac{4}{9}\right)\right]^2 = \left(-\dfrac{2}{9}\right)^2$.

$\left(x - \dfrac{2}{9}\right)^2 = -\dfrac{5}{81}$ Factor the left side and simplify the right side. A real number squared cannot be negative.

No real solution.

53. $3y^2 + 8y - 2 = 0$

$y^2 + \dfrac{8}{3}y - \dfrac{2}{3} = 0$ Divide by 3 so the coefficient of y^2 is 1.

$y^2 + \dfrac{8}{3}y = \dfrac{2}{3}$ Add $\dfrac{2}{3}$ to both sides.

$y^2 + \dfrac{8}{3}y + \dfrac{16}{9} = \dfrac{2}{3} + \dfrac{16}{9}$ Add $\dfrac{16}{9}$ to both sides: $\dfrac{16}{9} = \left[\dfrac{1}{2}\left(\dfrac{8}{3}\right)\right]^2 = \left(\dfrac{4}{3}\right)^2$.

$\left(y + \dfrac{4}{3}\right)^2 = \dfrac{22}{9}$ Factor the left side and simplify the right side.

$y + \dfrac{4}{3} = \pm\dfrac{\sqrt{22}}{3}$ Square root property.

Exercises 9.3

$y = -\dfrac{4}{3} \pm \dfrac{\sqrt{22}}{3}$ Solve for y.

$y = \dfrac{-4 \pm \sqrt{22}}{3}$ Simplify.

$\left\{\dfrac{-4 \pm \sqrt{22}}{3}\right\}$

57.

	Time	Fraction of the Pool Filled in One Hour
Fill the pool (drain shut)	t	$\dfrac{1}{t}$
Drain the pool (supply off)	$t + 2$	$\dfrac{1}{t+2}$
Supply on and drain open	8	$\dfrac{1}{8}$

$\begin{pmatrix}\text{Fraction of pool that is}\\ \text{filled in 1 hr with drain closed}\end{pmatrix} - \begin{pmatrix}\text{fraction of pool that is}\\ \text{drained in 1 hr with supply off}\end{pmatrix} = \begin{pmatrix}\text{fraction of pool that can be}\\ \text{filled in 1 hr with both open}\end{pmatrix}$

$\dfrac{1}{t} - \dfrac{1}{t+2} = \dfrac{1}{8}$ Fill in information from chart.

$8t(t+2)\left(\dfrac{1}{t} - \dfrac{1}{t+2}\right) = 8t(t+2)\left(\dfrac{1}{8}\right)$ Multiply both sides by the LCM.

$8t + 16 - 8t = t^2 + 2t$ Simplify.
$16 = t^2 + 2t$
$1 + 16 = t^2 + 2t + 1$ Add 1 to both sides: $1 = \left[\dfrac{1}{2}(2)\right]^2 = (1)^2$.

$17 = (t+1)^2$ Simplify the left side and factor the right side.
$\pm\sqrt{17} = t + 1$ Square root property.
$-1 \pm \sqrt{17} = t$ Solve for t.

Since $t > 0$, $t = -1 + \sqrt{17} \approx 3.1$ hr.

61.

	Distance /	Rate =	Time
Rana	7	r	$7/r$
Ben	8	$r + 5$	$8/(r+5)$

$\begin{pmatrix}\text{Rana's}\\ \text{time}\end{pmatrix} + \begin{pmatrix}\text{Ben's}\\ \text{time}\end{pmatrix} = \begin{pmatrix}\text{Total}\\ \text{time}\end{pmatrix}$

$$\frac{7}{r} + \frac{8}{r+5} = 1\frac{1}{2}$$ Fill in information from the chart.

$$2r(r+5)\left(\frac{7}{r} + \frac{8}{r+5}\right) = 2r(r+5)\left(\frac{3}{2}\right)$$ Multiply both sides by the LCM.

$$14r + 70 + 16r = 3r^2 + 15r$$ Simplify.
$$30r + 70 = 3r^2 + 15r$$
$$70 = 3r^2 - 15r$$

$$\frac{70}{3} = r^2 - 5r$$ Divide both sides by 3 so the coefficient of r^2 is 1.

$$\frac{25}{4} + \frac{70}{3} = r^2 - 5r + \frac{25}{4}$$ Add $\frac{25}{4}$ to both sides: $\frac{25}{4} = \left[\frac{1}{2}(-5)\right]^2 = \left(-\frac{5}{2}\right)^2$.

$$\frac{355}{12} = \left(r - \frac{5}{2}\right)^2$$ Simplify the left side and factor the right side.

$$\pm\frac{\sqrt{355}}{2\sqrt{3}} = r - \frac{5}{2}$$ Square root property.

$$\frac{5}{2} \pm \frac{\sqrt{355}}{2\sqrt{3}} = r$$ Solve for r.

Since $r > 0$, $r = \frac{5}{2} + \frac{\sqrt{355}}{2\sqrt{3}} \approx 7.9$ mph

and $r + 5 = 12.9$ mph.

65. One possible answer: To solve $ax^2 + bx + c = 0$ by completing the square follow these steps:

    1.) If $a \neq 1$, divide both sides by a so the coefficient of x^2 is 1. We want the coefficient to equal 1 because we are trying to get the left side in the form $r^2 + 2rs + s^2$.

    2.) Add $-\frac{c}{a}$ to both sides of the equation.

    3.) In the form we are striving for, $r^2 + 2rs + s^2$, the coefficient of r is $2s$. In order to decide the value of s^2, take half of $2s$ and square it. So in our equation, $\left[\frac{1}{2}\left(\frac{b}{a}\right)\right]^2 = \left(\frac{b}{2a}\right)^2 = \frac{b^2}{4a^2}$, is the value to be added to both sides of the equation.

    4.) Factor the perfect-square trinomial on the left side and simplify the right side.

Exercises 9.4

5.) Use the square root property to solve the equation that is now in the form $(x + d)^2 = e$. If $e < 0$, there is no real solution.

73. $(\sqrt{5} - \sqrt{2})^2 = (\sqrt{5})^2 - 2(\sqrt{5})(\sqrt{2}) + (\sqrt{2})^2$ Multiply using the formula $(a - b)^2 = a^2 - 2ab + b^2$.

$\qquad = 5 - 2\sqrt{10} + 2$ Simplify.

$\qquad = 7 - 2\sqrt{10}$

77. $\dfrac{\sqrt{3}}{2 - \sqrt{3}} = \dfrac{\sqrt{3}}{2 - \sqrt{3}} \cdot \dfrac{2 + \sqrt{3}}{2 + \sqrt{3}}$ Multiply the numerator and the denominator by the conjugate of the denominator.

$\qquad = \dfrac{2\sqrt{3} + 3}{2^2 - (\sqrt{3})^2}$ Multiply.

$\qquad = \dfrac{2\sqrt{3} + 3}{4 - 3}$ Simplify.

$\qquad = \dfrac{2\sqrt{3} + 3}{1}$

$\qquad = 2\sqrt{3} + 3$

EXERCISES 9.4

1. $x^2 - 3x - 18 = 0$ The equation is in standard form and $a = 1$, $b = -3$ and $c = -18$.

$x = \dfrac{-b \pm \sqrt{b^2 - 4ac}}{2a}$

$x = \dfrac{-(-3) \pm \sqrt{(-3)^2 - 4(1)(-18)}}{2(1)}$ Substitute these values into the formula.

$x = \dfrac{3 \pm \sqrt{9 + 72}}{2}$ Simplify.

$x = \dfrac{3 \pm \sqrt{81}}{2}$

$x = \dfrac{3 \pm 9}{2}$

$x = \dfrac{3 + 9}{2}$ or $x = \dfrac{3 - 9}{2}$

$x = 6$ or $x = -3$

$\{-3, 6\}$

5. $b^2 + 14b + 24 = 0$ The equation is in standard form and $a = 1$, $b = 14$ and $c = 24$.

$$x = \frac{-b \pm \sqrt{b^2 - 4ac}}{2a}$$

$$x = \frac{-14 \pm \sqrt{14^2 - 4(1)(24)}}{2(1)}$$

Substitute these values into the formula.

$$x = \frac{-14 \pm \sqrt{196 - 96}}{2}$$

Simplify.

$$x = \frac{-14 \pm \sqrt{100}}{2}$$

$$x = \frac{-14 \pm 10}{2}$$

$$x = \frac{-14 + 10}{2} \quad \text{or} \quad x = \frac{-14 - 10}{2}$$

$$x = -2 \quad \text{or} \quad x = -12$$

$\{-12, -2\}$

9. $x^2 - 10x - 24 = 0$ The equation is in standard form and $a = 1$, $b = -10$ and $c = -24$.

$$x = \frac{-b \pm \sqrt{b^2 - 4ac}}{2a}$$

$$x = \frac{-(-10) \pm \sqrt{(-10)^2 - 4(1)(-24)}}{2(1)}$$

Substitute these values into the formula.

$$x = \frac{10 \pm \sqrt{100 + 96}}{2}$$

Simplify.

$$x = \frac{10 \pm \sqrt{196}}{2}$$

$$x = \frac{10 \pm 14}{2}$$

$$x = \frac{10 + 14}{2} \quad \text{or} \quad x = \frac{10 - 14}{2}$$

$$x = 12 \quad \text{or} \quad x = -2$$

$\{-2, 12\}$

13. $x^2 + 9x - 70 = 0$ The equation is in standard form and $a = 1$, $b = 9$ and $c = -70$.

$$x = \frac{-b \pm \sqrt{b^2 - 4ac}}{2a}$$

$$x = \frac{-9 \pm \sqrt{9^2 - 4(1)(-70)}}{2(1)}$$

Substitute these values into the formula.

Exercises 9.4

$$x = \frac{-9 \pm \sqrt{81 + 280}}{2}$$ Simplify.

$$x = \frac{-9 \pm \sqrt{361}}{2}$$

$$x = \frac{-9 \pm 19}{2}$$

$$x = \frac{-9 + 19}{2} \quad \text{or} \quad x = \frac{-9 - 19}{2}$$

$$x = 5 \quad \text{or} \quad x = -14$$

$$\{-14, 5\}$$

17. $6w^2 - 5w - 6 = 0$ The equation is in standard form and $a = 6$, $b = -5$ and $c = -6$.

$$w = \frac{-b \pm \sqrt{b^2 - 4ac}}{2a}$$

$$w = \frac{-(-5) \pm \sqrt{(-5)^2 - 4(6)(-6)}}{2(6)}$$ Substitute these values into the formula.

$$w = \frac{5 \pm \sqrt{25 + 144}}{12}$$ Simplify.

$$w = \frac{5 \pm \sqrt{169}}{12}$$

$$w = \frac{5 \pm 13}{12}$$

$$w = \frac{5 + 13}{12} \quad \text{or} \quad w = \frac{5 - 13}{12}$$

$$w = \frac{3}{2} \quad \text{or} \quad w = -\frac{2}{3}$$

$$\left\{-\frac{2}{3}, \frac{3}{2}\right\}$$

21. $\quad 20x^2 - 23x = 21$ Rewrite in standard form.
$\quad 20x^2 - 23x - 21 = 0$ This is standard form so $a = 20$, $b = -23$ and $c = -21$.

$$x = \frac{-b \pm \sqrt{b^2 - 4ac}}{2a}$$

$$x = \frac{-(-23) \pm \sqrt{(-23)^2 - 4(20)(-21)}}{2(20)}$$ Substitute these values into the formula.

$$x = 23 \pm \sqrt{529 + 1680}$$ Simplify.

$$x = \frac{23 \pm \sqrt{2209}}{40}$$

$$x = \frac{23 \pm 47}{40}$$

$$x = \frac{23 + 47}{40} \quad \text{or} \quad x = \frac{23 - 47}{40}$$

$$x = \frac{7}{4} \quad \text{or} \quad x = -\frac{3}{5}$$

$$\left\{-\frac{3}{5}, \frac{7}{4}\right\}$$

25. $\quad 3x^2 - 4x = 7$ Rewrite in standard form.
$\quad\;\; 3x^2 - 4x - 7 = 0$ This is standard form with
$\qquad\qquad\qquad\qquad$ $a = 3$, $b = -4$ and $c = -7$.

$$x = \frac{-b \pm \sqrt{b^2 - 4ac}}{2a}$$

$$x = \frac{-(-4) \pm \sqrt{(-4)^2 - 4(3)(-7)}}{2(3)}$$ Substitute these values into the formula.

$$x = \frac{4 \pm \sqrt{15 + 84}}{6}$$ Simplify.

$$x = \frac{4 \pm \sqrt{100}}{6}$$

$$x = \frac{4 \pm 10}{6}$$

$$x = \frac{4 + 10}{6} \quad \text{or} \quad x = \frac{4 - 10}{6}$$

$$x = \frac{7}{3} \quad \text{or} \quad x = -1$$

$$\left\{-1, \frac{7}{3}\right\}$$

29. $\;-3t^2 - 8t + 3 = 0$ The equation is in standard form and
$\qquad\qquad\qquad\qquad$ $a = -3$, $b = -8$ and $c = 3$.

$$t = \frac{-b \pm \sqrt{b^2 - 4ac}}{2a}$$

$$t = \frac{-(-8) \pm \sqrt{(-8)^2 - 4(-3)(3)}}{2(-3)}$$ Substitute these values into the formula.

$$t = \frac{8 \pm \sqrt{64 + 36}}{-6}$$ Simplify.

Exercises 9.4

$$t = \frac{8 \pm \sqrt{100}}{-6}$$

$$t = \frac{8 \pm 10}{-6}$$

$$t = \frac{8 + 10}{-6} \quad \text{or} \quad t = \frac{8 - 10}{-6}$$

$$t = -3 \quad \text{or} \quad t = \frac{1}{3}$$

$$\left\{-3, \frac{1}{3}\right\}$$

33. $6x^2 + 22x = 21x + 77$ Rewrite in standard form.
$6x^2 + x - 77 = 0$ This is standard form and
     $a = 6$, $b = 1$ and $c = -77$.

$$x = \frac{-b \pm \sqrt{b^2 - 4ac}}{2a}$$

$$x = \frac{-1 \pm \sqrt{1^2 - 4(6)(-77)}}{2(6)}$$ Substitute these values into the formula.

$$x = \frac{-1 \pm \sqrt{1 + 1848}}{12}$$ Simplify.

$$x = \frac{-1 \pm \sqrt{1849}}{12}$$

$$x = \frac{-1 \pm 43}{12}$$

$$x = \frac{-1 + 43}{12} \quad \text{or} \quad x = \frac{-1 - 43}{12}$$

$$x = \frac{7}{2} \quad \text{or} \quad x = -\frac{11}{3}$$

$$\left\{-\frac{11}{3}, \frac{7}{2}\right\}$$

37. $x^2 + 10 = 5x$ Rewrite in standard form.
$x^2 - 5x + 10 = 0$ This is standard form and
     $a = 1$, $b = -5$ and $c = 10$.

$$x = \frac{-b \pm \sqrt{b^2 - 4ac}}{2a}$$

$$x = \frac{-(-5) \pm \sqrt{(-5)^2 - 4(1)(10)}}{2(1)}$$ Substitute these values into the formula.

$$x = \frac{5 \pm \sqrt{25 - 40}}{2}$$ Simplify.

$$x = \frac{5 \pm \sqrt{-15}}{2}$$ $\sqrt{-15}$ is not a real number.

No real solution

41. $4x^2 - 4x - 5 = 0$ This is standard form and $a = 4$, $b = -4$ and $c = -5$.

$$x = \frac{-b \pm \sqrt{b^2 - 4ac}}{2a}$$

$$x = \frac{-(-4) \pm \sqrt{(-4)^2 - 4(4)(-5)}}{2(4)}$$ Substitute these values into the formula.

$$x = \frac{4 \pm \sqrt{16 + 80}}{8}$$ Simplify.

$$x = \frac{4 \pm \sqrt{96}}{8}$$

$$x = \frac{4 \pm 4\sqrt{6}}{8}$$

$$x = \frac{4(1 \pm \sqrt{6})}{8}$$ Factor 4 out of the numerator.

$$x = \frac{1 \pm \sqrt{6}}{2}$$ Reduce.

$$\left\{ \frac{1 \pm \sqrt{6}}{2} \right\}$$

45. $(x + 3)^2 + 2x(x + 3) = 18$
$x^2 + 6x + 9 + 2x^2 + 6x = 18$ Multiply to clear the parentheses.
$3x^2 + 12x + 9 = 18$
$3x^2 + 12x - 9 = 0$
$x^2 + 4x - 3 = 0$ Divide both sides of the equation by 3. This is standard form and $a = 1$, $b = 4$ and $c = -3$.

$$x = \frac{-b \pm \sqrt{b^2 - 4ac}}{2a}$$

$$x = \frac{-4 \pm \sqrt{4^2 - 4(1)(-3)}}{2(1)}$$ Substitute these values into the formula.

$$x = \frac{-4 \pm \sqrt{16 + 12}}{2}$$ Simplify.

$$x = \frac{-4 \pm \sqrt{28}}{2}$$

Exercises 9.4

$$x = \frac{-4 \pm 2\sqrt{7}}{2}$$

$$x = \frac{2(-2 \pm \sqrt{7})}{2}$$ Factor 2 out of the numerator.

$$x = -2 \pm \sqrt{7}$$ Reduce.

$$\{-2 \pm \sqrt{7}\}$$

49. $2x^2 - 5x + 1 = 0$ This is in standard form and $a = 2$, $b = -5$ and $c = 1$.

$$x = \frac{-b \pm \sqrt{b^2 - 4ac}}{2a}$$

$$x = \frac{-(-5) \pm \sqrt{(-5)^2 - 4(2)(1)}}{2(2)}$$ Substitute these values into the formula.

$$x = \frac{5 \pm \sqrt{25 - 8}}{4}$$ Simplify.

$$x = \frac{5 \pm \sqrt{17}}{4}$$

$$x = \frac{5 + \sqrt{17}}{4} \quad \text{or} \quad x = \frac{5 - \sqrt{17}}{4}$$

$$x \approx 2.3 \quad \text{or} \quad x \approx 0.2$$

$$\{0.2, 2.3\}$$

53. $9a^2 + a - 2 = 0$ This is in standard form and $a = 9$, $b = 1$ and $c = -2$.

$$a = \frac{-b \pm \sqrt{b^2 - 4ac}}{2a}$$

$$a = \frac{-1 \pm \sqrt{1^2 - 4(9)(-2)}}{2(9)}$$ Substitute these values into the formula.

$$a = \frac{-1 \pm \sqrt{1 + 72}}{18}$$ Simplify.

$$a = \frac{-1 \pm \sqrt{73}}{18}$$

$$a = \frac{-1 + \sqrt{73}}{18} \quad \text{or} \quad a = \frac{-1 - \sqrt{73}}{18}$$

$$a \approx 0.4 \quad \text{or} \quad a \approx -0.5$$

$$\{-0.5, 0.4\}$$

57. $\quad s = 6 + 150t - 16t^2$ Formula.
$\quad\;\; 220 = 6 + 150t - 16t^2$ Replace s with 220.
$\quad\quad\; 0 = -214 + 150t - 16t^2$ Rewrite in standard form.
$\quad\quad\quad\quad\quad\quad\quad\quad\quad\quad\quad\quad\;$ $a = -16$, $b = 150$ and $c = -214$.

$$t = \frac{-b \pm \sqrt{b^2 - 4ac}}{2a}$$

$$t = \frac{-150 \pm \sqrt{150^2 - 4(-16)(-214)}}{2(-16)}$$ Substitute these values into the formula.

$$t = \frac{-150 \pm \sqrt{22500 - 13696}}{-32}$$ Simplify.

$$t = \frac{-150 \pm \sqrt{8804}}{-32}$$

$$t = \frac{-150 + \sqrt{8804}}{-32} \quad \text{or} \quad t = \frac{-150 - \sqrt{8804}}{-32}$$

$t = 1.76 \quad\quad\quad\;\; \text{or} \quad t = 7.62$

1.76 sec, or 7.62 sec

61. $\quad i = t^2 - 10t + 20$ Formula.
$\quad\; 12 = t^2 - 10t + 20$ Replace i with 12.
$\quad\quad 0 = t^2 - 10t + 8$ Rewrite in standard form.
$\quad\quad\quad\quad\quad\quad\quad\quad\quad\quad\quad\quad$ $a = 1$, $b = -10$ and $c = 8$.

$$t = \frac{-b \pm \sqrt{b^2 - 4ac}}{2a}$$

$$t = \frac{-(-10) \pm \sqrt{(-10)^2 - 4(1)(8)}}{2(1)}$$ Substitute these values into the formula.

$$t = \frac{10 \pm \sqrt{100 - 32}}{2}$$ Simplify.

$$t = \frac{10 \pm \sqrt{68}}{2}$$

$$t = \frac{10 + \sqrt{68}}{2} \quad \text{or} \quad t = \frac{10 - \sqrt{68}}{2}$$

$t = 9.12 \quad\quad\quad \text{or} \quad t = 0.88$

0.88 sec, or 9.12 sec

65. One possible answer:

$\quad 2x^2 - 19x + 9 = 0$
$\quad (2x - 1)(x - 9) = 0$
$\quad 2x = 1 \quad \text{or} \quad x - 9 = 0$
$\quad 2x = 1 \quad \text{or} \quad\quad\; x = 9$
$\quad\;\; x = \dfrac{1}{2}$

Exercises 9.4

$2x^2 - 19x + 9 = 0$

$x = \dfrac{-b \pm \sqrt{b^2 - 4ac}}{2a}$

$x = \dfrac{-(-19) \pm \sqrt{(-19)^2 - 4(2)(9)}}{2(2)}$

$x = \dfrac{19 \pm \sqrt{361 - 72}}{4}$

$x = \dfrac{19 \pm \sqrt{289}}{4}$

$x = \dfrac{19 \pm 17}{4}$

$x = \dfrac{19 + 17}{4}$ or $x = \dfrac{19 - 17}{4}$

$x = 9$ or $x = \dfrac{1}{2}$

$6x^2 - 25x + 25 = 0$
$(2x - 5)(3x - 5) = 0$

$2x - 5 = 0$ or $3x - 5 = 0$
$\quad 2x = 5$ or $\quad 3x = 5$

$\quad x = \dfrac{5}{2}$ or $\quad x = \dfrac{5}{3}$

$6x^2 - 25x + 25 = 0$

$x = \dfrac{-b \pm \sqrt{b^2 - 4ac}}{2a}$

$x = \dfrac{-(-25) \pm \sqrt{(-25)^2 - 4(6)(25)}}{2(6)}$

$x = \dfrac{25 \pm \sqrt{625 - 600}}{12}$

$x = \dfrac{25 \pm \sqrt{25}}{12}$

$x = \dfrac{25 \pm 5}{12}$

$x = \dfrac{25 + 5}{12}$ or $x = \dfrac{25 - 5}{12}$

$x = \dfrac{5}{2}$ or $x = \dfrac{5}{3}$

$$x^2 + \sqrt{3}x - \sqrt{2}x - \sqrt{6} = 0$$
$$x(x + \sqrt{3}) - \sqrt{2}(x + \sqrt{3}) = 0$$
$$(x + \sqrt{3})(x - \sqrt{2}) = 0$$

$$x + \sqrt{3} = 0 \quad \text{or} \quad x - \sqrt{2} = 0$$
$$x = -\sqrt{3} \quad \text{or} \quad x = \sqrt{2}$$
$$x \approx -1.73 \quad \text{or} \quad x \approx 1.41$$

$$x = \frac{-b \pm \sqrt{b^2 - 4ac}}{2a}$$

$$x = \frac{-(\sqrt{3} - \sqrt{2}) \pm \sqrt{(\sqrt{3} - \sqrt{2})^2 - 4(1)(-\sqrt{6})}}{2(1)}$$

$$x = \frac{-\sqrt{3} + \sqrt{2} \pm \sqrt{3 - 2\sqrt{6} + 2 + 4\sqrt{6}}}{2}$$

$$x = \frac{-\sqrt{3} + \sqrt{2} \pm \sqrt{5 + 2\sqrt{6}}}{2}$$

$$x = \frac{-\sqrt{3} + \sqrt{2} + \sqrt{5 + 2\sqrt{6}}}{2} \quad \text{or} \quad x = \frac{-\sqrt{3} + \sqrt{2} - \sqrt{5 + 2\sqrt{6}}}{2}$$

$$x \approx 1.41 \quad \text{or} \quad x \approx -1.73$$

Factoring is preferable unless the coefficients are very large and it is not easy to factor.

73. $15a^3b + 10a^2b^3 - 15a^2bc^2$
   $= 5a^2b(3a + 2b^2 - 3c^2)$ Factor out the GCF $5a^2b$.

77. $\sqrt{27} - \sqrt{75} + \sqrt{300}$
   $= 3\sqrt{3} - 5\sqrt{3} + 10\sqrt{3}$ Simplify $\sqrt{27}$, $\sqrt{75}$ and $\sqrt{300}$.
   $= (3 - 5 + 10)\sqrt{3}$ Distributive property.
   $= 8\sqrt{3}$

EXERCISES 9.5

1. $x^2 - 4x - 5 = 0$
   $(x - 5)(x + 1) = 0$ Factor.
   $x - 5 = 0 \quad \text{or} \quad x + 1 = 0$ Zero-product property.
     $x = 5 \quad \text{or} \quad x = -1$
   $\{-1, 5\}$

5. $3x^2 - x - 9 = 0$ This is in standard form and
                $a = 3$, $b = -1$, and $c = -9$.

   $x = \dfrac{-b \pm \sqrt{b^2 - 4ac}}{2a}$

   $x = \dfrac{-(-1) \pm \sqrt{(-1)^2 - 4(3)(-9)}}{2(3)}$ Substitute these values into the formula.

Exercises 9.5

$$x = \frac{1 \pm \sqrt{1 + 108}}{6}$$ Simplify.

$$x = \frac{1 \pm \sqrt{109}}{6}$$

$$\left\{\frac{1 \pm \sqrt{109}}{6}\right\}$$

9. $x(x - 5) = 5(4 - x)$
    $x^2 - 5x = 20 - 5x$ Multiply to clear the parentheses.
    $x^2 = 20$ Add $5x$ to both sides.
    $x = \pm 2\sqrt{5}$ Square root property.
    $\{\pm 2\sqrt{5}\}$

13. $x(x + 5) = 3(5 - x)$
    $x^2 + 5x = 15 - 3x$ Multiply to clear the parentheses.
    $x^2 + 8x - 15 = 0$ Write in standard form.
    $a = 1, b = 8$ and $c = -15$.

$$x = \frac{-b \pm \sqrt{b^2 - 4ac}}{2a}$$

$$x = \frac{-8 \pm \sqrt{8^2 - 4(1)(-15)}}{2(1)}$$ Substitute these values into the formula.

$$x = \frac{-8 \pm \sqrt{64 + 60}}{2}$$ Simplify.

$$x = \frac{-8 \pm \sqrt{124}}{2}$$

$$x = \frac{-8 \pm 2\sqrt{31}}{2}$$

$$x = \frac{2(-4 \pm \sqrt{31})}{2}$$ Factor 2 out of the numerator.

$x = -4 \pm \sqrt{31}$ Reduce.

$\{-4 \pm \sqrt{31}\}$

17. $x(2x - 1) + 3x(1 - 4x) + 8 = 0$
    $2x^2 - x + 3x - 12x^2 + 8 = 0$ Multiply to clear the parentheses.
    $-10x^2 + 2x + 8 = 0$ Combine like terms.
    $5x^2 - x - 4 = 0$ Divide both sides by -2.
    $(5x + 4)(x - 1) = 0$ Factor the left side.
    $5x + 4 = 0$ or $x - 1 = 0$ Zero-product property.
    $5x = -4$ or $x = 1$
    $x = -\dfrac{4}{5}$

$\left\{-\dfrac{4}{5}, 1\right\}$

21. $3x(x - 6) + 2 = (2x - 4)(x - 7) + 3$ Multiply to clear the parentheses.
$3x^2 - 18x + 2 = 2x^2 - 18x + 28 + 3$
$3x^2 - 18x + 2 = 2x^2 - 18x + 31$
$x^2 = 29$ Write in the form $x^2 = K$.
$x = \pm\sqrt{29}$ Square root property.
$\{\pm\sqrt{29}\}$

25. $3 + \dfrac{2}{x} - \dfrac{5}{x^2} = 0$

$x^2\left(3 + \dfrac{2}{x} - \dfrac{5}{x^2}\right) = x^2(0)$ Multiply both sides by x^2 to clear the fractions.

$3x^2 + 2x - 5 = 0$
$(3x + 5)(x - 1) = 0$ Factor the left side.

$3x + 5 = 0$ or $x - 1 = 0$ Zero-product property.
$3x = -5$ or $x = 1$

$x = -\dfrac{5}{3}$

$\left\{-\dfrac{5}{3}, 1\right\}$

29. $2 + \dfrac{11}{x} + \dfrac{5}{x^2} = 0$

$x^2\left(2 + \dfrac{11}{x} + \dfrac{5}{x^2}\right) = x^2(0)$ Multiply both sides by x^2 to clear the fractions.

$2x^2 + 11x + 5 = 0$
$(2x + 1)(x + 5) = 0$ Factor the left side.

$2x + 1 = 0$ or $x + 5 = 0$ Zero-product property.
$2x = -1$ or $x = -5$

$x = -\dfrac{1}{2}$

$\left\{-5, -\dfrac{1}{2}\right\}$

33. $\dfrac{12}{x - 1} = 10 - \dfrac{12}{x}$

$x(x - 1)\left(\dfrac{12}{x - 1}\right) = x(x - 1)\left(10 - \dfrac{12}{x}\right)$ Multiply both sides by $x(x - 1)$ to clear the fractions.

$12x = 10x(x - 1) - 12(x - 1)$
$12x = 10x^2 - 10x - 12x + 12$
$12x = 10x^2 - 22x + 12$
$0 = 10x^2 - 34x + 12$ Write in standard form.
$0 = 5x^2 - 17x + 6$ Divide both sides by 2.
$0 = (5x - 2)(x - 3)$ Factor the right side.

Exercises 9.5

$$5x - 2 = 0 \quad \text{or} \quad x - 3 = 0$$
$$5x = 2 \quad \text{or} \quad x = 3$$
$$x = \frac{2}{5}$$

$$\left\{\frac{2}{5}, 3\right\}$$

37. $\quad (2x + 3)^2 = 21 + 10x$
$x^2 + 6x + 9 = 21 + 10x$ Multiply to clear parentheses.
$x^2 - 4x - 12 = 0$ Write in standard form.
$(x - 6)(x + 2) = 0$ Factor the left side.

$x - 6 = 0 \quad \text{or} \quad x + 2 = 0$ Zero-product property.
$x = 6 \quad \text{or} \quad x = -2$

$\{-2, 6\}$

41. $\quad (2x + 3)^2 = 53 - x^2$
$4x^2 + 12x + 9 = 53 - x^2$ Multiply to clear parentheses.
$5x^2 + 12x - 44 = 0$ Write in standard form.
$(5x + 22)(x - 2) = 0$ Factor the left side.

$5x + 22 = 0 \quad \text{or} \quad x - 2 = 0$ Zero-product property.
$5x = -22 \quad \text{or} \quad x = 2$
$x = -\frac{22}{5}$

$\left\{-\frac{22}{5}, 2\right\}$

45. $\quad (x - 5)(x + 2) = 139 - (2x - 3)(x + 4)$
$x^2 - 3x - 10 = 139 - (2x^2 + 5x - 12)$ Multiply.
$x^2 - 3x - 10 = 139 - 2x^2 - 5x + 12$
$x^2 - 3x - 10 = -2x^2 - 5x + 151$
$3x^2 + 2x - 161 = 0$ Write in standard form.
$(3x + 23)(x - 7) = 0$ Factor the left side.

$3x + 23 = 0 \quad \text{or} \quad x - 7 = 0$ Zero-product property.
$3x = -23 \quad \text{or} \quad x = 7$
$x = -\frac{23}{3}$

$\left\{-\frac{23}{3}, 7\right\}$

49. $\quad \dfrac{7}{x + 3} - 1 = \dfrac{5}{x - 2}$

$(x + 3)(x - 2)\left(\dfrac{7}{x + 3} - 1\right) = (x + 3)(x - 2)\left(\dfrac{5}{x - 2}\right)$ Multiply both sides by $(x + 3)(x - 2)$ to clear the fractions.

$7(x - 2) - (x + 3)(x - 2) = 5(x + 3)$ Simplify.

$$7x - 14 - (x^2 + x - 6) = 5x + 15 \qquad \text{Simplify.}$$
$$7x - 14 - x^2 - x + 6 = 5x + 15$$
$$-x^2 + 6x - 8 = 5x + 15$$
$$0 = x^2 - x + 23 \qquad \text{Write in standard form.}$$
$$\qquad\qquad\qquad\qquad\qquad a = 1, \ b = -1 \text{ and } c = 23.$$

$$x = \frac{-b \pm \sqrt{b^2 - 4ac}}{2a}$$

$$x = \frac{-(-1) \pm \sqrt{(-1)^2 - 4(1)(23)}}{2(1)} \qquad \text{Substitution these values into the formula.}$$

$$x = \frac{1 \pm \sqrt{1 - 92}}{2} \qquad \text{Simplify.}$$

$$x = \frac{1 \pm \sqrt{-91}}{2} \qquad \sqrt{-91} \text{ is not a real number.}$$

No real solution

53. $\quad x - \dfrac{9x}{x - 2} = \dfrac{-10}{x - 2}$

$$(x - 2)\left(x - \frac{9x}{x - 2}\right) = (x - 2)\left(\frac{-10}{x - 2}\right) \qquad \text{Multiply both sides by } x - 2 \text{ to clear the fractions.}$$

$$x^2 - 2x - 9x = -10 \qquad \text{Simplify.}$$
$$x^2 - 11x = -10$$
$$x^2 - 11x + 10 = 0 \qquad \text{Write in standard form.}$$
$$(x - 10)(x - 1) = 0 \qquad \text{Factor the left side.}$$

$$x - 10 = 0 \quad \text{or} \quad x - 1 = 0 \qquad \text{Zero-product property.}$$
$$x = 10 \quad \text{or} \quad x = 1$$

$\{1, 10\}$

57.

$$(14 - 2x)(18 - 2x) = 140 \qquad A = wl.$$
$$252 - 64x + 4x^2 = 140$$
$$4x^2 - 64x + 112 = 0 \qquad \text{Write in standard form.}$$
$$x^2 - 16x + 28 = 0 \qquad \text{Divide both sides by 4.}$$
$$(x - 14)(x - 2) = 0 \qquad \text{Factor the left side.}$$

$$x - 14 = 0 \quad \text{or} \quad x - 2 = 0 \qquad \text{Zero-product property.}$$
$$x = 14 \quad \text{or} \quad x = 2$$

Since $14 - 2x$ and $18 - 2x$ must be positive, $x = 2$ ft.

Exercises 9.5

61. $r = \dfrac{b^2 + 4h^2}{8h}$ Formula.

$30 = \dfrac{24^2 + 4h^2}{8h}$ Replace r with 30 and b with 24.

$240h = 24^2 + 4h^2$ Multiply both sides by $8h$.
$0 = 4h^2 - 240h + 576$ Write in standard form.
$0 = h^2 - 60h + 144$ Divide both sides by 4.

$h = \dfrac{-b \pm \sqrt{b^2 - 4ac}}{2a}$ $a = 1, b = -60, c = 144$.

$h = \dfrac{-(-60) \pm \sqrt{(-60)^2 - 4(1)(144)}}{2(1)}$

$h = \dfrac{60 \pm \sqrt{3600 - 576}}{2}$

$h = \dfrac{60 \pm \sqrt{3024}}{2}$

$h = 60 + \dfrac{\sqrt{3024}}{2}$ or $h = \dfrac{60 - \sqrt{3024}}{2}$

$h = 57.5$ or $h = 2.5$

Since $h < r$, $h = 2.5$ ft.

65. car's rate: $r + 10$
    bus' rate: r

In 3 hours, distance traveled by:
car: $3(r + 10) = 3r + 30$
bus: $3r$

$(3r)^2 + (3r + 30)^2 = 180^2$ Pythagorean formula.
$9r^2 + 9r^2 + 180r + 900 = 32400$
$18r^2 + 180r + 900 = 32400$
$18r^2 + 180r - 31500 = 0$ Write in standard form.
$r^2 + 10r - 1750 = 0$ Divide both sides by 18.

$$r = \frac{-b \pm \sqrt{b^2 - 4ac}}{2a}$$
$a = 1, b = 10, c = -1750.$

$$r = \frac{-10 \pm \sqrt{10^2 - 4(1)(-1750)}}{2(1)}$$

$$r = \frac{-10 \pm \sqrt{100 + 7000}}{2}$$

$$r = \frac{-10 \pm \sqrt{7100}}{2}$$

$$r = \frac{-10 + \sqrt{7100}}{2} \quad \text{or} \quad r = \frac{-10 - \sqrt{7100}}{2}$$

$r = 37.1 \quad \text{or} \quad r = -47.1$

Since $r > 0$, $r = 37.1$.
car's rate: $r + 10 = 37.1 + 10 = 47.1$ mph.
bus' rate: $r = 37.1$ mph.

69. One possible answer: It is best to use the quadratic formula when the quadratic equation cannot be factored; when $a \neq 1$ and b is odd; and when $b \neq 0$.

73. $\{5 \pm \sqrt{10}\}$

$x = 5 + \sqrt{10}$	or	$x = 5 - \sqrt{10}$
$x - (5 + \sqrt{10}) = 0$	or	$x - (5 - \sqrt{10}) = 0$
$x - 5 - \sqrt{10} = 0$	or	$x - 5 + \sqrt{10} = 0$

Set x equal to each solution.
Get 0 on one side of the equation.

$$(x - 5 - \sqrt{10})(x - 5 + \sqrt{10}) = 0$$
$$[(x - 5) - \sqrt{10}][(x - 5) + \sqrt{10}] = 0$$
$$(x - 5)^2 - (\sqrt{10})^2 = 0$$
$$x^2 - 10x + 25 - 10 = 0$$
$$x^2 - 10x + 15 = 0$$

If one of the factors is 0 then the product is 0.
Write the factors in conjugate form.
Multiply.

77. $3(x + 4) - 3(5 - 2x) - 4(3x + 9) = 2(x + 6)$
$3x + 12 - 15 + 6x - 12x - 36 = 2x + 12$
$-3x - 39 = 2x + 12$
$-39 = 5x + 12$
$-51 = 5x$

Multiply.
Combine like terms.
Add $3x$ to both sides.
Subtract 12 from both sides.

$$-\frac{51}{5} = x$$

Divide both sides by 5.

$\left\{-\dfrac{51}{5}\right\}$

81. $\sqrt{2x + 5} - 9 = -12$
$\sqrt{2x + 5} = -3$
No real solution

Add 9 to both sides. A square root cannot equal a negative number.

EXERCISES 9.6

1. $y = x^2 - 2$

x	y
-2	$(-2)^2 - 2 = 2$
-1	$(-1)^2 - 2 = -1$
0	$0^2 - 2 = -2$
1	$1^2 - 2 = -1$
2	$2^2 - 2 = 2$

Find some solution pairs.

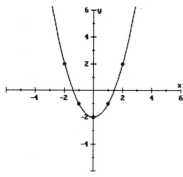

Plot the (x, y) points and connect with a smooth curve.

5. $y = x^2 + 6$

x	y
-2	$(-2)^2 + 6 = 10$
-1	$(-1)^2 + 6 = 7$
0	$0^2 + 6 = 6$
1	$1^2 + 6 = 7$
2	$2^2 + 6 = 10$

Find some solution pairs.

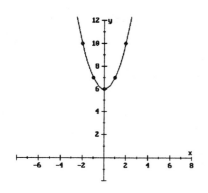

Plot the (x, y) points and connect with a smooth curve.

9. $y = x^2 + 2x$

x	y
-3	$(-3)^2 + 2(-3) = 3$
-2	$(-2)^2 + 2(-2) = 0$
-1	$(-1)^2 + 2(-1) = -1$
0	$0^2 + 2(0) = 0$
1	$1^2 + 2(1) = 3$

Find some solution pairs.

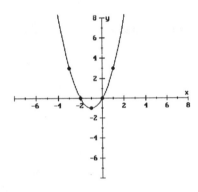

Plot the (x, y) points and connect with a smooth curve.

13. $y = x^2 - 5x$

x	y
0	$0^2 - 5(0) = 0$
1	$1^2 - 5(1) = -4$
2.5	$(2.5)^2 - 5(2.5) = -6.25$
4	$4^2 - 5(4) = -4$
5	$5^2 - 5(5) = 0$

Find some solution pairs.

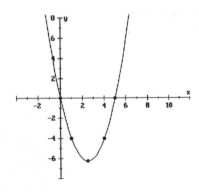

Plot the (x, y) points and connect with a smooth curve.

Exercises 9.6

17. $y = x^2 - 2x + 2$

x	y
-1	$(-1)^2 - 2(-1) + 2 = 5$
0	$0^2 - 2(0) + 2 = 2$
1	$1^2 - 2(1) + 2 = 1$
2	$2^2 - 2(2) + 2 = 2$
3	$3^2 - 2(3) + 2 = 5$

Find some solution pairs.

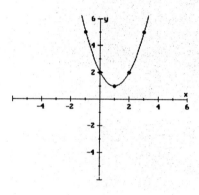

Plot the (x, y) points and connect with a smooth curve.

21. $y = 0.5x^2 - 3$

x	y
-2	$0.5(-2)^2 - 3 = -1$
-1	$0.5(-1)^2 - 3 = -2.5$
0	$0.5(0)^2 - 3 = -3$
1	$0.5(1)^2 - 3 = -2.5$
2	$0.5(2)^2 - 3 = -1$

Find some solution pairs.

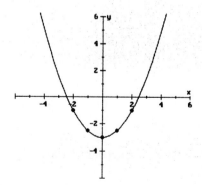

Plot the (x, y) points and connect with a smooth curve.

25. $s = 180t - 16t^2$

x	y
0	$180(0) - 16(0)^2 = 0$
2	$180(2) - 16(2)^2 = 296$
5	$180(5) - 16(5)^2 = 500$
7	$180(7) - 16(7)^2 = 476$
10	$180(10) - 16(10)^2 = 200$

Find some solution pairs.

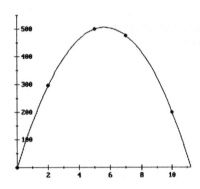

Plot the (x, y) points and connect with a smooth curve.

$s = 180(6) - 16(6)^2$
$s = 504$ ft

Replace t with 6.

29. One possible answer: When the quadratic equation is in the form $y = ax^2 + bx + c$ you can determine how the parabola opens by the sign of a. If $a > 0$, the parabola opens upward and if $a < 0$, the parabola opens downward.

33. $\quad x^2 - 2 = 3$
$\qquad x^2 = 5$
$\qquad x = \pm\sqrt{5}$

$\{\pm\sqrt{5}\}$

Write in the form $x^2 = K$.
Square root property.

37. $\quad \sqrt{x - 2} = 3$
$\quad (\sqrt{x - 2})^2 = 3^2$
$\qquad x - 2 = 9$
$\qquad\qquad x = 11$

Square both sides.
Solve for x.

Check:

$\sqrt{11 - 2} = 3$
$\quad \sqrt{9} = 3$ is true.

$\{11\}$

Substitute $x = 11$ into the original equation.

41. $\sqrt{x} + \sqrt{2} = \sqrt{3}$
$\sqrt{x} = \sqrt{3} - \sqrt{2}$ Isolate the variable.
$(\sqrt{x})^2 = (\sqrt{3} - \sqrt{2})^2$ Square both sides.
$x = 3 - 2\sqrt{6} + 2$
$x = 5 - 2\sqrt{6}$ Simplify.

$\{5 - 2\sqrt{6}\}$

CHAPTER 9 CONCEPT REVIEW

1. True

2. True

3. True

4. False;   If $A \cdot B = 0$, then $A = 0$ or $B = 0$.

5. True

6. False; $a^2 + b^2 = c^2$ where c is the hypotenuse.

7. True

8. False;   You must first divide both sides of the equation by 2 so the coefficient of x^2 is 1. Next divide the coefficient of x by 2 and square that number.

9. True

10. True

CHAPTER 9 TEST

1. $x^2 + 16x - 8 = 0$
$x^2 + 16x = 8$ Add 8 to both sides.
$x^2 + 16x + 64 = 8 + 64$ Add 64 to both sides: $64 = \left[\frac{1}{2}(16)\right]^2 = (8)^2$.
$(x + 8)^2 = 72$ Factor the left side and simplify the right side.
$x + 8 = \pm 6\sqrt{2}$ Square root property.
$x = -8 \pm 6\sqrt{2}$ Solve for x.

$\{-8 \pm 6\sqrt{2}\}$

2. $12x^2 = 768$
$x^2 = 64$ Divide both sides by 12.
$x = \pm 8$ Square root property.

$\{\pm 8\}$

3. $a^2 + b^2 = c^2$ Pythagorean formula.
   $6^2 + 7^2 = c^2$ Replace a with 6 and b with 7.
   $36 + 49 = c^2$
   $85 = c^2$
   $9.2 = c$

4. $2x^2 + x + 7 = 0$ The equation is in standard form and $a = 2$, $b = 1$, and $c = 7$.

   $x = \dfrac{-b \pm \sqrt{b^2 - 4ac}}{2a}$

   $x = \dfrac{-1 \pm \sqrt{1^2 - 4(2)(7)}}{2(2)}$ Substitute these values into the formula.

   $x = \dfrac{-1 \pm \sqrt{1 - 56}}{4}$

   $x = \dfrac{-1 \pm \sqrt{-55}}{4}$ $\sqrt{-55}$ is not a real number.

   No real solution

5. $x^2 - 6x - 14 = 0$
   $x^2 - 6x = 14$ Add 14 to both sides.
   $x^2 - 6x + 9 = 14 + 9$ Add 9 to both sides: $9 = \left[\dfrac{1}{2}(-6)\right]^2 = (-3)^2$.
   $(x - 3)^2 = 23$ Factor the left side and simplify the right side.
   $x - 3 = \pm\sqrt{23}$ Square root property.
   $x = 3 \pm \sqrt{23}$

   $\{3 \pm \sqrt{23}\}$

6. $3x^2 - 3x - 2 = 0$ The equation is in standard form and $a = 3$, $b = -3$ and $c = -2$.

   $x = \dfrac{-b \pm \sqrt{b^2 - 4ac}}{2a}$

   $x = \dfrac{(-3) \pm \sqrt{(-3)^2 - 4(3)(-2)}}{2(3)}$ Substitute these values into the formula.

   $x = \dfrac{3 \pm \sqrt{9 + 24}}{6}$ Simplify.

   $x = \dfrac{3 \pm \sqrt{33}}{6}$

   $\left\{\dfrac{3 \pm \sqrt{33}}{6}\right\}$

7. $3x^2 - 17 = 2x^2 + 15$
   $x^2 = 32$ Write in the form $x^2 = K$.
   $x = \pm 4\sqrt{2}$ Square root property.
   $\{\pm 4\sqrt{2}\}$

Chapter 9 Test

8. $x^2 + 4x - 17 = 0$ The equation is in standard form and $a = 1$, $b = 4$ and $c = -17$.

$x = \dfrac{-b \pm \sqrt{b^2 - 4ac}}{2a}$

$x = \dfrac{-4 \pm \sqrt{4^2 - 4(1)(-17)}}{2(1)}$ Substitute these values into the formula.

$x = \dfrac{-4 \pm \sqrt{16 + 68}}{2}$ Simplify.

$x = \dfrac{-4 \pm \sqrt{84}}{2}$

$x = \dfrac{-4 \pm 2\sqrt{21}}{2}$

$x = \dfrac{2(-2 \pm \sqrt{21})}{2}$ Factor the numerator.

$x = -2 \pm \sqrt{21}$

$\{-2 \pm \sqrt{21}\}$

9. $(x - 4)(2x + 3) = x(x - 4) + 8$
   $2x^2 - 5x - 12 = x^2 - 4x + 8$ Multiply to clear the parentheses.
       $x^2 - x - 20 = 0$ Write in standard form.
      $(x - 5)(x + 4) = 0$ Factor the left side.

   $x - 5 = 0$ or $x + 4 = 0$ Zero-product property.
       $x = 5$ or $x = -4$

$\{-4, 5\}$

10.

[Right triangle with legs $175 - x$ (vertical) and x (horizontal), and hypotenuse 125.]

$x^2 + (175 - x)^2 = 125^2$ Pythagorean formula.
$x^2 + 30625 - 350 + x^2 = 15625$
$2x^2 - 350x + 30625 = 15625$
$2x^2 - 350x + 15000 = 0$ Write in standard form.
$x^2 - 175x + 7500 = 0$ Divide both sides by 2.
$(x - 75)(x - 100) = 0$ Factor the left side.
$x - 75 = 0$ or $x - 100 = 0$ Zero-product property.
$x = 75$ or $x = 100$
$175 - x = 100$ or $175 - x = 75$

The dimensions are 75 yd by 100 yd.